The Driving Test

first time.

The Driving Test

PASS

first time.

by

Linda Hatswell B Ed (Hons)

 The Driving School

Published by AA Publishing (a trading name of
Automobile Association Developments Limited,
whose registered office is Norfolk House,
Priestley Road, Basingstoke,
Hampshire RG24 9NY.
Registered Number 1878835.)

A catalogue record for this book is available
from the British Library

ISBN 0 7495 0719 5

Edited by Susan Gordon

Printed in Spain

Foreword

You want to pass the driving test and take advantage of the freedom and mobility that driving a car can give you. Do the following three things and you will achieve your objective — passing the test:

1

Learn and understand the **skills** of driving by taking lessons from a trained and fully qualified driving instructor.

2

Acquire the **knowledge** of the rules through your instructor and by studying *The Highway Code*. A key element of learning is to test and reinforce your knowledge. This book is specially designed for this purpose.

3

Take the right **attitude**. No one is a 'natural' or a 'perfect driver'. All drivers make mistakes. Be careful, courteous and considerate to all other road users.

Using this book shows you have the right attitude to learning to drive. So, remember, acquire the **skills**, the **knowledge** and the right **attitude** and you will pass the test!

Nick Bravery, Business manager

 **The Driving School**

6 **Contents**

Contents 7

Part 9

Part 10

Part 11

Part 12

Part 13

Part 14

Part 15

Part 16

Part 17

About this book

This book is aimed at those who are learning or are about to learn how to drive.

No book can teach you how to drive. This is best left to an Approved Driving Instructor who has proven skill and ability. This book will, however, help you test your knowledge of the theoretical aspects of driving.

A thorough knowledge of driving theory is becoming increasingly important, especially in view of the harmonisation of driving tests throughout the European Community states, due to take place by 1996. Most EC countries have a knowledge or theory test as well as a practical test of driving ability.

This book is designed to be easy to follow and is made up of questions, many of which use diagrams and illustrations, that test your knowledge of driving on today's roads. We have included a section on the driving test itself: what do you really know about taking the test and what is expected of you? And of course learning to drive does not stop with passing the test. It continues throughout life, and the new experiences and challenges it will bring such as driving on the Continent and dealing with a road accident are also covered here.

How to use this book

It is strongly recommended that you learn to drive with an Approved Driving Instructor. All AA The Driving School's franchised instructors are fully qualified and have passed the AA's own strict test of instructional ability. The Driving School offers a programme of planned tuition which includes frequent checks to keep you up to date with your own progress, ensuring that you are fully prepared and ready to pass your test.

The book is divided into clearly defined sections. Each deals with a particular requirement of the test, for example reverse parallel parking, aspects of car control or road procedure. A recognised syllabus is followed, each heading being indentifiable as a subject that needs to be covered when learning to drive. The questions are colour coded, matching the answers which are to be found towards the back of the book.

The best method of working is to read the question, write in your answer, then check what you have written against the answer we have given. If you need further explanation, we suggest you refer to the HMSO publications *Driving, Your Driving Test* and *The Highway Code*, all of which are available from AA shops.

The book is written so that it can be used either in conjunction with taking driving lessons, at home to test your knowledge of the subjects covered on your lessons, or between lessons to review the work covered.

With the increase in traffic and today's complex systems, it is not enough to be able to drive a car. You need to know that you fully understand all road procedures, traffic signs — and much more. This book will test how much knowledge you really have and will help you to be a safer driver on the road.

Good defensive driving depends on adopting the right attitude from the start. These questions will test your knowledge of what is required before you even sit in the driver's seat.

Q1 What do you need before you can drive on a public road?

Answer _Current signed Full clean licence_

Q2 The best way to learn is to have regular planned tuition with an ADI (Approved Driving Instructor).

An ADI is someone who has taken and passed all three driving instructor's e*examinations* and is on the official r*egister*

Complete the sentence

Q3 Anyone supervising a learner must be at least _21_ years old and must have held (and still hold) a full licence (motor car) for at least t*hree* years.

Complete the sentence

Q4 Your tuition vehicle must display L-plates.

Where should they be placed?

Answer _Front and Back, not obstructing the windows_

ANSWERS ON PAGE 112

Q5 Young and inexperienced drivers are more vulnerable.

Is this true or false? Answer _True_

Q6 Showing responsibility to yourself and others is the key to being a safe driver.

Ask yourself, would you . . .

YES NO

☐ ☒ 1 Want to drive with someone who has been drinking?

☐ ☒ 2 Want to drive with someone who takes risks and puts other lives at risk?

☐ ☒ 3 Want to drive with someone who does not concentrate?

☐ ☒ 4 Want to drive with someone who drives too fast?

Q7 Do you want to be a safe and responsible driver?

YES NO

☒ ☐ Tick the appropriate box

ANSWERS ON PAGE 112

Q8   To use the controls safely you need to adopt a suitable driving position. There are a number of checks you should make.

Fill in the missing words

1   Check the hANDBRAKE is on.

2   Check the dOORS are shut.

3   Check your sEAT is in the correct position.

4   Check the hEAD rESTRAINS is adjusted to give maximum protection.

5   Check the driving mIRRORS are adjusted to give maximum rear view.

6   Check your sEAT bELT is securely fastened.

ANSWERS ON PAGE 112

Q9 Here is a list of controls and a list of functions.

Match each control to its function by placing the appropriate letter in the box

The controls	The functions
`E` The handbrake	A To control the direction in which you want to travel
`D` The driving mirrors	B To slow or stop the vehicle
`F` The gear lever	C To increase or decrease the engine's speed
`G` The clutch	D To give you a clear view behind
`A` The steering wheel	E To hold the vehicle still when it is stationary
`B` The foot-brake	F To enable you to change gear
`C` The accelerator	G To enable you to make or break contact between the engine and the wheels

Fill in the missing word   The accelerator can also be called the g*AS* pedal.

Q10 Which foot should you use for each of these controls (in cars with a manual gearbox)?

R = Right foot   L = Left foot

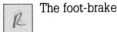 `R` The foot-brake   `L` The clutch   `R` The accelerator

Q11 Are the following statements about steering true or false?

TRUE   FALSE

[ ] [F] 1 I must keep both hands on the wheel at all times.

[T] [ ] 2 To keep good control I should feed the wheel through my hands.

[ ] [F] 3 I can place my hands at any position as long as I am comfortable.

[ ] [F] 4 When going round corners, it is best to cross my hands (hand over hand).

[T] [ ] 5 I should never take both hands off the wheel when the vehicle is moving.

[T] [ ] 6 To straighten up I should feed the wheel back through my hands.

Q12 Match each of the following controls to its function.

The controls

[B] The direction indicators

[A] Dipped beam

[D] Main beam

[C] Rear fog lamp

[E] Horn

[F] Hazard lights

The functions

A To enable you to see the road ahead and other road users to see you without causing dazzle

B To show other road users which way you intend to turn

C To use only when visibility is 100m/yds or less

D To enable you to see further, but not to be used when there is oncoming traffic

E To warn other road users of your presence

F To warn other road users when you are temporarily obstructing traffic

Q1 The following is a list of actions involved in moving off from rest.

Number the boxes 1 to 9 to show the correct sequence

The first box has been filled in to give you a start

1   A Press the clutch down fully.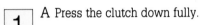

5   B Check your mirrors.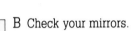

3   C Set the gas (accelerator) pedal.

2   D Move the gear lever into 1st gear.

6   E Decide if you need to give a signal.

4   F Let the clutch come up to biting point and hold it steady.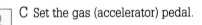

7   G Check your blind spot.

8   H If safe, release the handbrake and let the clutch up a little more.

9   I Press the gas pedal a little more and let the clutch up fully.

ANSWERS ON PAGE 112

Q2   The following is a list of actions required for stopping normally. Number the boxes 1 to 9 to show the correct sequence. The first box has been filled in to give you a start

1 A Check your mirrors.

3 B Take your foot off the gas (accelerator) pedal.

2 C Decide if you need to signal and, if necessary, do so.

4 D Press the brake pedal, lightly at first and then more firmly.

6 E As the car stops, ease the pressure off the foot-brake (except when you are on a slope).

5 F Just before the car stops, press the clutch pedal right down.

8 G Put the gear lever into neutral.

7 H Apply the handbrake fully.

9 I Take both feet off the pedals.

ANSWERS ON PAGE 113

Gears enable you to select the power you need from the engine to perform

a particular task.

Q3 Which gear gives you the most power?

Answer 1st

Q4 If you were travelling at 60 mph on a clear road,

which gear would you most likely select?

Answer 4/5 th

Q5 When approaching and

turning a corner, as shown in the

diagram, which gear would you

most likely use?

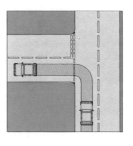

Answer 2nd

Q6 You need to change gear to match your e _NG I NE_ speed to

the speed at which your v _ECHICE_ is travelling. The s _OUND_ the

engine is making will help you know w _HEN_ to change gear.

Complete the sentences

ANSWERS ON PAGE 113

Q7 Number the boxes to show the correct sequence of actions required when changing up.

The first box has been filled in for you

1 A Place your left hand on the gear lever.
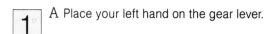

3 B Move the gear lever to the next highest position.
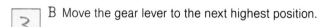

2 C Press the clutch pedal down fully and ease off the gas pedal.

4 D Let the clutch pedal come up fully and, at the same time, press the gas pedal.

5 E Put your left hand back on the steering wheel.

ANSWERS ON PAGE 113

Q8   Are the following statements about changing down true or false?

Tick the appropriate boxes

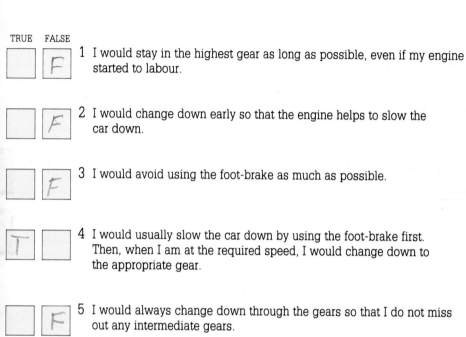

TRUE   FALSE

[ ] [F] 1 I would stay in the highest gear as long as possible, even if my engine started to labour.

[ ] [F] 2 I would change down early so that the engine helps to slow the car down.

[ ] [F] 3 I would avoid using the foot-brake as much as possible.

[T] [ ] 4 I would usually slow the car down by using the foot-brake first. Then, when I am at the required speed, I would change down to the appropriate gear.

[ ] [F] 5 I would always change down through the gears so that I do not miss out any intermediate gears.

ANSWERS ON PAGE 113

Q9 When changing gear, I should look . . .

1 Ahead   2 At the gear lever   3 At my feet

1 Which is correct?

ANSWERS ON PAGE 113

Q10 Do's and don'ts

DO	DON'T	
	✓	1 Force the gear lever if there is any resistance
	✓	2 Rush the gear changes
✓		3 Match your speed with the correct gear
✓		4 Use the brakes, where necessary, to reduce speed before changing down
✓		5 Listen to the sound of the engine
	✓	6 Take your eyes off the road when changing gear
	✓	7 Hold the gear lever longer than necessary
	✓	8 Coast with the clutch down or the gear lever in neutral

Q11
Which wheels turn when you turn the steering wheel?

A The front wheels

B The back wheels

Answer

Q12
When you turn your steering wheel to the right, which way do your wheels turn?

A To the right

B To the left

Answer

Q13
The steering lock is . . .

A The locking mechanism that stops the steering wheel from moving when the ignition key is removed

B The angle through which the wheels turn when the steering wheel is turned

Answer

Q14
Which wheels follow the shorter pathway?

A The front wheels

B The back wheels

Answer

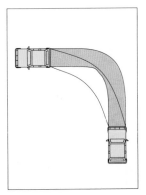

ANSWERS ON PAGE 113

Q15 Which is the correct position for normal driving?

Put letter A, B or C in the box

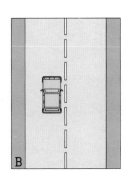
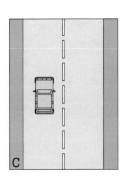

Answer C

Q16 Which diagram shows the correct pathway when driving normally?

Put a letter A or B in the box

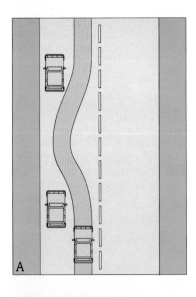
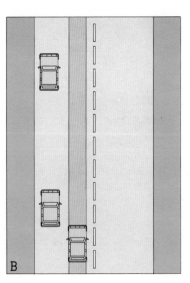

Answer B

ANSWERS ON PAGE 113

Q17 Pushing the clutch pedal down . . .

A Releases the engine from the wheels

B Engages the engine with the wheels   Answer

Q18 The point where the clutch plates meet is called the

bITIING point.

Fill in the missing word

Q19 By controlling the amount of contact between the clutch plates,

it is possible to control the speed of the car. Would you use this control . . .

YES   NO

Y [ ]   1 When moving away from rest?

Y [ ]   2 When manoeuvring the car in reverse gear?

[ ] N   3 When slowing down to turn a corner?

Y [ ]   4 In very slow moving traffic?

[ ] N   5 To slow the car down?

ANSWERS ON PAGE 114

Q1 A junction is a point where t<u>WO</u> o<u>R</u> m<u>ORE</u> r<u>OADS</u> meet.

Complete the sentence

Q2 Here are five main types of junction. Name them

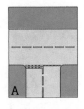 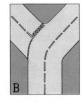  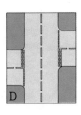 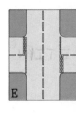

T JUNCTION   Y JUNCTION   ROUNDABOUT   STAGGARD   CROSSROAD

Q3 Match these road signs to the junctions shown above.

Put letters A, B, C, D and E in the boxes

1 E   2 C   3 B   4 D   5 A

Q4 What do these signs mean?

1 Stop and give way

2 Slow down, look, and proceed if safe

3 Give way to traffic on the major road

A

1 Answer

Put a number in each box

Answer 3
B

Q5   At every junction you should follow a safe routine.

Put the following into the correct order by numbering the boxes 1 to 5

2 Signal   4 Speed   3 Position   1 Mirrors   5 Look

Q6   The diagram shows a car turning right into a minor road. The boxes are numbered to show the correct sequence of actions.

Complete the sentence

At point 5 you should look and a s s E s s the situation, d e c i d e to go or wait, and a c t accordingly.

ANSWERS ON PAGE 114

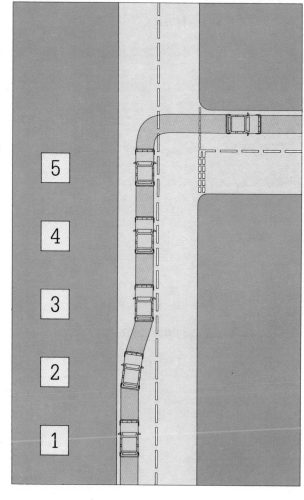

Q7 Turning left into a minor road

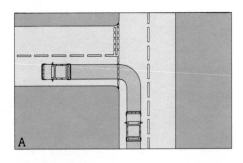

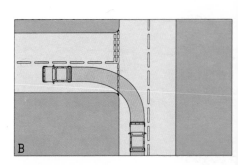

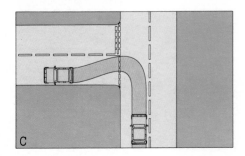

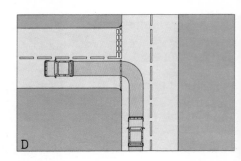

Which diagram shows the best path to follow when driving a motor car A, B, C or D?

Answer

Q8 You turn into a side road. Pedestrians are already crossing it.

Should you ...   Tick the appropriate box

☐ A Sound your horn

☐ B Slow down and give way

☐ C Flash your lights

☐ D Wave them across

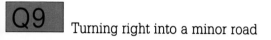

Q9 Turning right into a minor road

ANSWERS ON PAGE 114

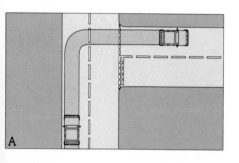

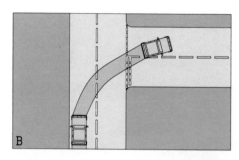

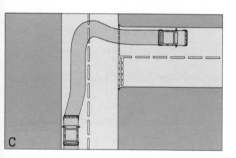

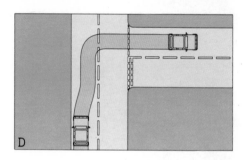

Which diagram shows the best path to follow A, B, C or D?

Answer

Q10 These are the golden rules for emerging from junctions.

Complete the sentences

1 Always use your m i RR o RS to check the speed and

  p o S i T i o N of vehicles behind.

2 Always cancel your s i G N A L.

3 Speed up to a s A FE speed after joining the new road.

4 Keep a s A FE d i S T A N CE between you and the vehicle ahead.

5 Do not attempt to o VE R T A KE until you can assess the new road.

Q11 All crossroads must be approached with caution.

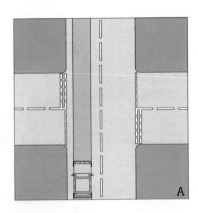

A   3   Action

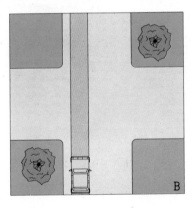

B   1   Action

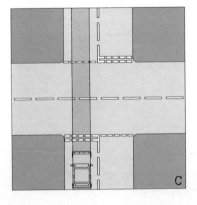

C   2   Action

Match Actions 1, 2 and 3 listed below with these diagrams

Actions

1 Approach with caution, look well ahead and be prepared to stop. Remember other drivers may assume they have priority.

2 Look well ahead, slow down and be prepared to give way to traffic on the major road.

3 Look well ahead and into the side roads for approaching vehicles. Remember other drivers may not give you priority.

ANSWERS ON PAGE 114

Q12 Which of the following statements describes the correct
procedure when approaching a roundabout? Put letter A, B or C in the box

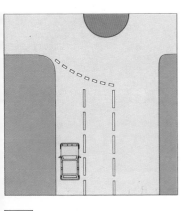

A The broken white line at a roundabout means
I must stop and give way to traffic already on the
roundabout.

B The broken white line at a roundabout means
I must give priority to traffic already on the
roundabout.

C The broken white line means I should give
way to any traffic approaching from my
immediate right.

C Answer

Q13 The following sentences give guidance on lane discipline on a
roundabout. Fill in the missing words

1 When turning left at a roundabout, I should stay in the L E F T -hand
lane and should stay in that lane throughout.

2 When going ahead at a roundabout, I should be in the L E F T -hand
lane, and should stay in that lane throughout unless conditions dictate
otherwise.

3 When turning right at a roundabout, I should approach in the
R I G H T -hand lane, or approach as if turning
right at a junction, and stay in that lane throughout.

ANSWERS ON PAGE 114

Q14 The letters A, B and C in the diagram mark places where you should signal.

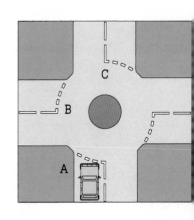

Complete the sentences

1 I would signal at A when turning LEFT.

2 I would signal at B when going AHEAD.

3 I would signal at A and at **C** when turning RIGHT.

Q15 At a roundabout you should always use a safe routine.

Fill in the missing words

mIRRORS, sIGNAL, pOSITION, sPEED, lOOK.

Q16 What does this sign mean?

1   Roundabout

2   Mini-roundabout

3   Vehicles may pass either side

Answer

ANSWERS ON PAGES 114-15

Q1 It is safest to park off the road or in a car park whenever possible. If you have to park on the road, think . . .   Fill in the missing words

1 Is it s_A_F_E_ ?   2 Is it c_o_N_U_L_E_N_A_N_I_ ?   3 Is it l_E_G_A_L_ ?

Q2 In this diagram four of the cars are parked illegally or without consideration of others.

Put the numbers of these cars in the boxes

| 1 | 2 | 3 | 6 |

For reverse parallel parking manoeuvres, see pages 52-3.

ANSWERS ON PAGE 115

Q1 The diagram shows a stationary vehicle on the left-hand side of the road. Which has priority, vehicle 1 or vehicle 2?

 Answer

ANSWER ON PAGE 115

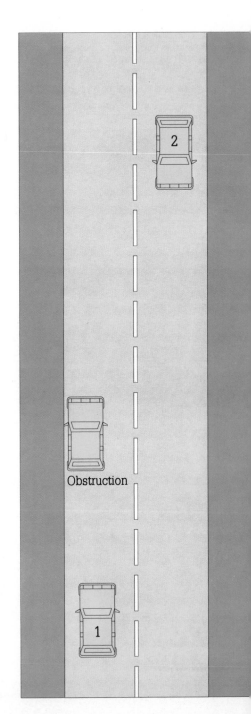

Obstruction

Q2 This diagram shows a steep downward hill with an obstruction on the right-hand side of the road.

Which vehicle should be given priority, vehicle 1 or vehicle 2?

 Answer

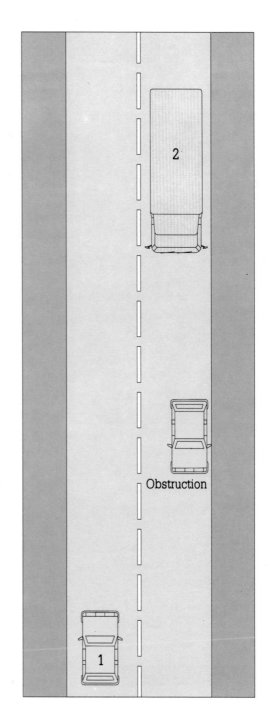

ANSWER ON PAGE 115

Q3   The diagram shows two vehicles, travelling in opposite directions, turning right at a crossroads.

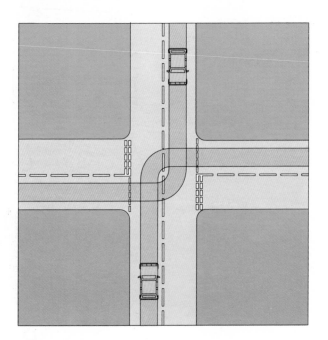

Are these statements true or false? Tick the appropriate boxes

TRUE   FALSE

☑ ☐   1 The safest route is to pass each other offside to offside.

☐ ☑   2 If the approaching vehicle flashes its headlamps, I should turn as quickly as possible.

☑ ☐   3 I should always try to get eye-to-eye contact with the driver of the other vehicle to determine which course to take.

ANSWERS ON PAGE 115

Q4 Which of the following factors, illustrated in the diagram, should be taken into consideration when turning right into a side road?

Tick the appropriate boxes

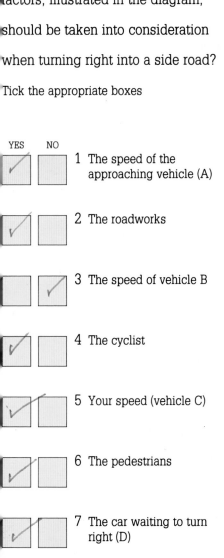

YES   NO

1 The speed of the approaching vehicle (A)

2 The roadworks

3 The speed of vehicle B

4 The cyclist

5 Your speed (vehicle C)

6 The pedestrians

7 The car waiting to turn right (D)

ANSWERS ON PAGE 115

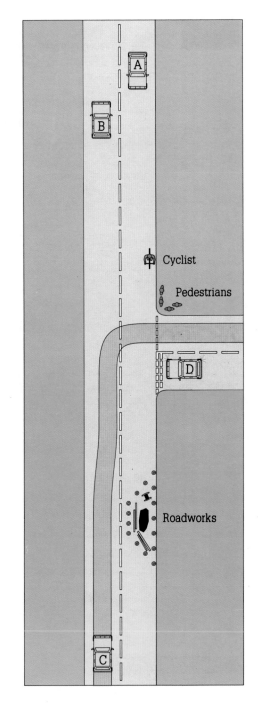

Q1 Are the following statements true or false when stopping in an emergency?

Tick the appropriate boxes

TRUE   FALSE

[✓] [ ] 1 Stopping in emergency increases the risk of skidding.

[✓] [ ] 2 I should push the brake pedal down harder as I slow down.

[✓] [ ] 3 It is important to react quickly.

[ ] [✓] 4 Putting the clutch pedal down helps to slow the car down.

[ ] [✓] 5 I should always remember to look in my mirrors as I slow down.

[ ] [✓] 6 I should signal left to tell other road users what I am doing.

[✓] [ ] 7 I should keep both hands on the wheel.

[✓] [ ] 8 I should always check my mirrors and look round before moving off.

ANSWERS ON PAGE 115

Q2 Cadence braking is a technique which can be used in very

slippery conditions in an emergency. Fill in the missing words

The technique requires you to p_u_m_p the brake pedal. The procedure to

follow is:

1 Apply m_AXAMINE pressure.

2 Release the brake pedal just as the wheels are about to l_ock .

3 Then q_uickly apply the brakes again. Apply and release the

brakes until vehicle has stopped. This technique should only be used in

emergency situations.

Q3 Anti-lock braking systems (ABS)* work in a similar way to

cadence braking. Fill in the missing words

When braking in an emergency, ABS brakes allow you to s_teer and

b_rake at the same time. You do not have to p_ump the brakes as

you would in cadence braking. When using ABS you keep the

p_ressure applied.

Are these statements about ABS braking true or false? Tick the correct boxes

TRUE FALSE

[ ] [F] 1 Cars fitted with ABS braking cannot skid.

ANSWERS ON PAGE 115

[ ] [F] 2 I do not need to leave as much room between me and the car in front if I have ABS brakes because I know I can stop in a shorter distance.

* ABS is a registered trade mark of Bosch (Germany). ABS stands for Anti-Blockiersystem

The distance taken for a car to reach stopping point divides into thinking distance and braking distance.

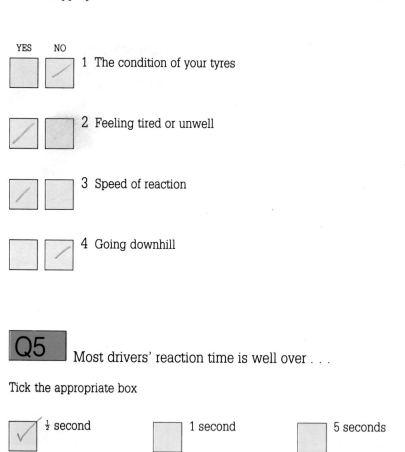

Q4 Could these factors affect thinking distance?

Tick the appropriate boxes

YES NO

1 The condition of your tyres

2 Feeling tired or unwell

3 Speed of reaction

4 Going downhill

Q5 Most drivers' reaction time is well over . . .

Tick the appropriate box

✓ ½ second   1 second   5 seconds

ANSWERS ON PAGE 116

Q6   Stopping distance depends partly on the speed at which the car is travelling.   Complete the sentences

1 At 30 mph your overall stopping distance will be 23 metres or 75 feet.

2 At 50 mph your thinking distance will be 15 metres or 50 feet.

3 At 70 mph your overall stopping distance will be 96 metres or 315 feet.

Q7   Stopping distance also varies according to road conditions.

Complete the sentences

In wet weather your vehicle will take lONGER to stop. You should therefore allow moRE time.

Q8   Too many accidents are caused by drivers driving too close to the vehicle in front. A safe gap between you and the vehicle in front can be measured by noting a stationary object and counting in seconds the time that lapses between the vehicle in front passing that object and your own vehicle passing that object.   Complete the sentence

Only a fool bREAKS the tWO sECOND rule.

ANSWERS ON PAGE 116

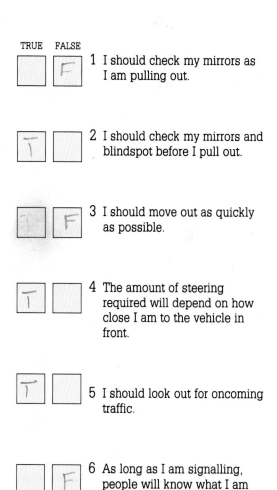

Q1 Are these statements about moving off at an angle true or false?

Tick the appropriate boxes

TRUE   FALSE

1 I should check my mirrors as I am pulling out. [F]

2 I should check my mirrors and blindspot before I pull out. [T]

3 I should move out as quickly as possible. [F]

4 The amount of steering required will depend on how close I am to the vehicle in front. [T]

5 I should look out for oncoming traffic. [T]

6 As long as I am signalling, people will know what I am doing. I will be able to pull out because somebody will let me in. [F]

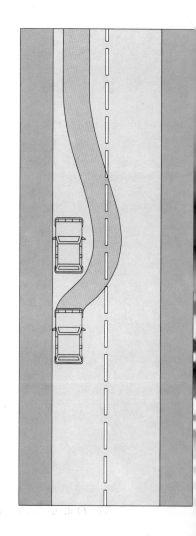

ANSWERS ON PAGE 116

Are these statements about moving off uphill true or false?

Tick the appropriate boxes

TRUE FALSE

1 On an uphill gradient the car will tend to roll back.

2 To stop the car rolling back I need to use more gas.

3 I do not need to use the handbrake.

4 The biting point may be slightly higher.

5 I need to press the gas pedal further than when moving off on the level.

6 I need to allow more time to pull away.

7 The main controls I use will be the clutch pedal, the gas pedal and the handbrake.

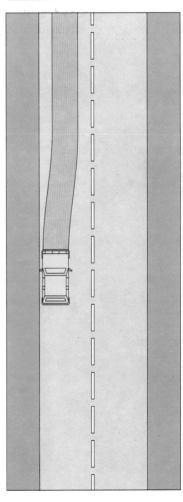

ANSWERS ON PAGE 116

Q3 Are these statements about moving off downhill true or false?

Tick the appropriate boxes

TRUE   FALSE

1 The car will tend to roll forwards.

2 The main controls I use will be the handbrake, the clutch pedal and the gas pedal.

3 The only gear I can move off in is 1st gear.

4 I should release the handbrake while keeping the foot-brake applied.

5 I should look round just before moving off.

6 I must not have my foot on the foot-brake as I start to release the clutch.

ANSWERS ON PAGE 116

Q4 The following statements are about approaching a junction when going uphill or downhill. With which do you agree? Tick the relevant boxes

When going downhill . . .

 1 It is more difficult to slow down

 2 Putting the clutch down will help slow the car down

 3 The higher the gear, the greater the control

 4 When changing gear you may need to use the foot-brake at the same time as the clutch

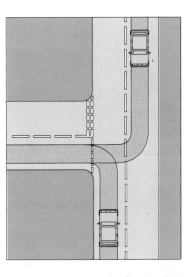

When going uphill . . .

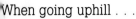 5 Early use of mirrors, signals, brakes, gears and steering will help to position the car correctly

 6 You may need to use your handbrake more often

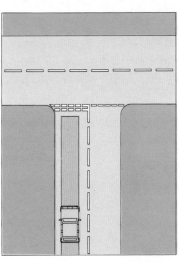

 7 When you change gear, the car tends to slow down

ANSWERS ON PAGE 116

Q1 Before reversing there are three things to consider.

Fill in the missing words

1 Is it s_A F E_ ?

2 Is it c_O N V E D A E N T_ ?

ANSWERS ON PAGES 116-17

3 Is it within the l_A W_ ?

Q2 Are the following statements about reversing true or false?

Tick the appropriate boxes

TRUE   FALSE

1 Other road users should see what I am doing and wait for me.

2 I should wave pedestrians on, so that I can get on with the manoeuvre more quickly.

3 I should avoid being too hesitant.

4 I should avoid making other road users slow down or change course.

Q3 How should you hold the steering wheel when reversing left?

A   B   C

Which is correct? A

Q4   These statements are all about reversing.

Tick those which you think are correct

[✓] 1 My car will respond differently in reverse gear.

[ ] 2 My car will feel no different.

[ ] 3 Steering is not affected. The car responds the same as when going forward.

[✓] 4 The steering will feel different. I will have to wait for the steering to take effect.

Q5   Which way will the rear of the car go when it is reversed?

Car A

Answer _LEFT_

Car B

Answer _RIGHT_

Q6   It is important to move the vehicle slowly when reversing.

Complete the sentence

Moving the vehicle slowly is safer because

I have control and it allows me to carry

out good oBSERVATION checks.

ANSWERS ON PAGE 117

Q7 When reversing good observation is vital. Where should you look?

Tick the correct answer

☐ 1 At the kerb   ☑ 3 Where your car is going

☐ 2 Ahead   ☐ 4 Out of the back window

Q8 Are these statements about reversing round a corner true or false?

Tick the appropriate boxes

TRUE　FALSE

☑ ☐ 1 If the corner is sharp, I need to be further away from the kerb.

☐ ☑ 2 The distance from the kerb makes no difference.

☑ ☐ 3 I should try to stay reasonably close to the kerb all the way round.

Q9 Before reversing I should check . . .　Tick the correct box

☐ 1 Behind me   ☐ 3 My door mirrors

☐ 2 Ahead and to the rear   ☑ 4 All round

ANSWERS ON PAGE 117

This diagram shows a car about to reverse round the corner to the left.

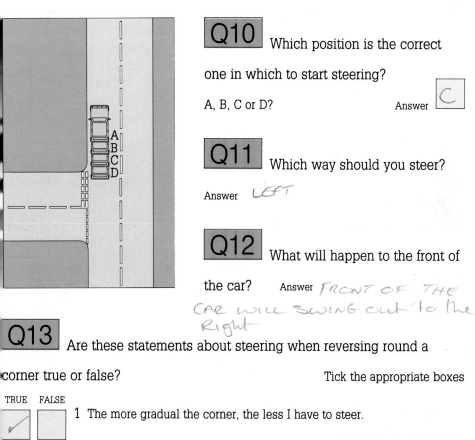

Q10 Which position is the correct one in which to start steering?

A, B, C or D?   Answer   C

Q11 Which way should you steer?

Answer   LEFT

Q12 What will happen to the front of the car?   Answer   FRONT OF THE CAR WILL SWING out to the Right

Q13 Are these statements about steering when reversing round a corner true or false?

Tick the appropriate boxes

TRUE   FALSE

1 The more gradual the corner, the less I have to steer.   [✓]

2 I need to steer the same for every corner.   [✓ FALSE]

ANSWERS ON PAGE 117

3 The sharper the corner, the more I have to steer.   [✓ TRUE]

Q14 As I enter the new road I should continue to keep a look-out for pedestrians and other road users. I should stop if necessary.

Complete the sentences

Q15 True or false? When reversing from a major road into a side road on the right, I have to move to the wrong side of the road.

TRUE   FALSE

[✓] [ ]   Tick the appropriate box

Q16 Which diagram shows the correct path to follow when moving to the right-hand side of the road?

A or B?

Answer

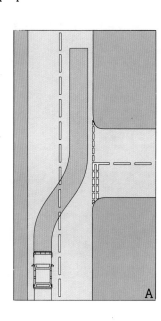

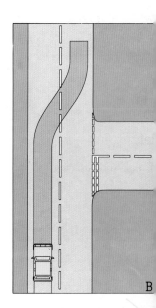

Q17 Which of the following correctly describes your sitting position for reversing to the right?

1 I will need to sit so that I can see over my right shoulder.

2 I will need to sit so that I can see over my right shoulder, ahead and to the left.

3 My position is the same as when reversing to the left.

Answer

ANSWERS ON PAGE 117

Q18 True or false? I may need to change my hand position on the wheel.

Tick the appropriate box

TRUE FALSE

Q19 True or false? It is easier to judge my position from the kerb when reversing to the right than when reversing to the left.

Tick the appropriate box

TRUE FALSE

Q20 Reversing to the right is more dangerous than reversing to the left because . . .

1 I cannot see as well

2 I am on the wrong side of the road

3 I might get in the way of vehicles emerging from the side road

Which statement is correct? 1, 2 or 3

Answer

Q21 How far down the side road should you reverse before moving over to the left-hand side?

Which diagram is correct? A or B?

Answer

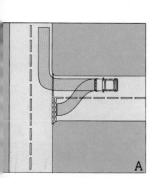

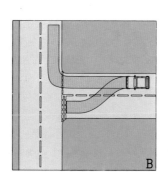

ANSWERS ON PAGE 117

Q22 Look at the diagrams and decide which is safer.

A Reversing into a side road   B Turning round in the road

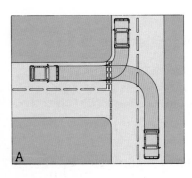 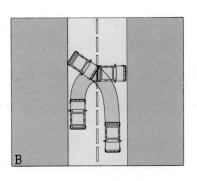

Answer A

Q23 The secret of turning in the road is to move the vehicle s_LOWLY and steer b_RISKLY .

Complete the sentence

Q24 True or false? I must be able to complete the manoeuvre in three moves:
1 forward, 2 reverse, 3 forward.

TRUE   FALSE

 Tick the appropriate box

Q25 Before manoeuvring, what should you take into consideration?

Tick the appropriate boxes

☐ 1 The size of your engine

☑ 2 The width of the road

☑ 3 The road camber

☑ 4 The steering circle of your vehicle

☐ 5 Parking restrictions

ANSWERS ON PAGE 117

Q26 Before moving forward, it is important to check a__L__L__ r_o_u_n_d_

for other road users.

Complete the sentence

Q27 Turning in the road requires proper use of the steering wheel.

Answer the following questions

1 When going forwards, which way should you steer?

Answer _RIGHT_

2 Before you reach the kerb ahead, what should you do?

Answer _BRISKLY LEFT_ (STEER)

3 When reversing, which way should you steer?

Answer _LEFT_

4 Before you reach the kerb behind you, what should you do?

Answer _STEER BRISKLY RIGHT_

5 As you move forward again, which way should you steer to straighten up?

Answer _RIGHT_

Q28 Reversing is a potentially dangerous manoeuvre. Good

observation is essential.   Answer the following questions

1 If you are steering left when reversing, which shoulder should you look over?

Answer _LEFT_

2 As you begin to steer to the right, where should you look?

Answer _OVER YOUR RIGHT_

ANSWERS ON PAGES 117-18

When parking between two cars . . .

1 The car is more manoeuvrable when driving forwards

2 The car is more manoeuvrable when reversing

3 There is no difference between going into the space forwards or reversing into it

Which statement is correct? 1, 2 or 3?   Answer

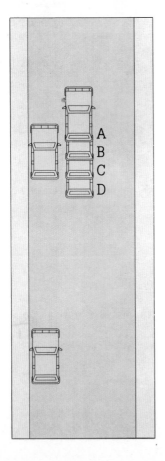

 The diagram shows a car preparing to reverse into a parking space. Which position is the correct one in which to start steering left?

Answer

Q31 With practice you should be able to park in a gap . . .

1 Your own car length

2 1½ times your own car length

3 2 times your own car length

4 2½ times your own car length

 Answer

ANSWERS ON PAGE 118

Q32
Use the diagram to help you answer the following questions.

1 Which way would you steer?

Answer _To the Left_

2 At this point what would you try to line up with the offside (right-hand side) of your vehicle?

Answer _The nearside headlight_

3 As you straighten up what do you have to be careful of?

Answer _Clipping the rear offside of lead car_

4 What do you need to do to straighten up?

Answer _Take off left lock_

5 What would you need to do in order to position the vehicle parallel to the kerb? Answer _Steer to the Right and take off the right lock._

Q33
True or false? During my driving test . . . Tick the appropriate boxes

	TRUE	FALSE	
1		✓	I will certainly be asked to perform this manoeuvre
2		✓	I have to be able to park between two cars
3	✓		Only the lead car will be present

ANSWERS ON PAGE 118

Q34
When carrying out this manoeuvre, where is it important to look?

Answer _ALL AROUND_

Q1   Traffic lights have three lights, red, amber, and green, which change from one to the other in a set order. Number the boxes 1 to 5 to show the correct order. The first one has been filled in to give you a start.

| 4 | Amber | 1 | Red | 2 | Red and amber | 5 | Red | 3 | Green |

Q2   What do the colours mean?

Fill in the correct colour for each of the following

1 Go ahead if the way is clear.   Colour GREEN

2 Stop and wait.   Colour RED/AMBER

3 Stop unless you have crossed the stop line or you are
   so close to it that stopping might cause an accident.   Colour AMBER

4 Stop and wait at the stop line.   Colour RED

Q3   Which of the following statements are true? Tick the appropriate boxes

On approach to traffic lights you should . . .

F   1 Speed up to get through before they change

T   2 Be ready to stop

ANSWERS ON PAGE 118

T   3 Look out for pedestrians

F   4 Sound your horn to urge pedestrians to cross quickly

Q4

Some traffic lights have green filters. Do they mean . . .

1 You can filter in the direction of the arrow only when the main light

is showing green?

2 You can filter even when the main light is not showing green?   Answer 2

Q5

The diagram shows the three lanes at a set of traffic lights.

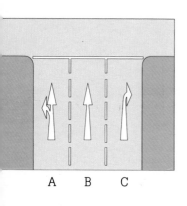

A   B   C

Which lane would you use for . . .

1 Going ahead   Answer _LANE A or B_

2 Turning right   Answer _LANE C_

3 Turning left   Answer _LANE A_

ANSWERS ON PAGE 118

Q6

At some traffic lights and junctions you will see yellow

criss-cross lines (box junctions). Can you . . .   Tick the appropriate boxes

YES   NO

☐ ✓ 1 Wait within them when going ahead if
        your exit is not clear?

☐ ✓ 2 Wait within them when going right if
        your exit is not clear?

✓ ☐ 3 Wait within them if there is oncoming
        traffic stopping you turning right but
        your exit is clear?

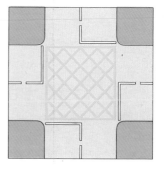

Pedestrians have certain rights of way at pedestrian crossings.

Q7 On approach to a zebra crossing, drivers will notice four features.

Name them

1 _ZIG ZAG lINES_

2 _Flashing yellow becons on both sides of the Road_

3 _Black and white stripes on the crossing_

4 _A give way line_

Q8 Are these statements about pedestrian crossings true or false?

TRUE FALSE

[✓] [ ] 1 I cannot park or wait on the zig-zag lines on the approach to a zebra crossing.

[✓] [ ] 2 I cannot park or wait on the zig-zag lines on either side of the crossing.

[ ] [✓] 3 I can overtake on the zig-zag lines on the approach to a crossing as long as the other vehicle is travelling slowly.

[✓] [ ] 4 I must give way to a pedestrian once he/she has stepped on to the crossing.

[✓] [ ] 5 If, on approach to a crossing, I intend to slow down or stop, I should use a slowing-down arm signal.

ANSWERS ON PAGES 118-19

Tick the appropriate boxes

Pelican crossings are light-controlled crossings.

Q9 On approach to a pelican crossing drivers will notice three key features.

Name them

1 _Traffice lights_

2 _Zig Zag Lines_

3 _A white stop line_

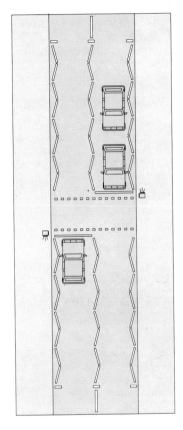

Q10 If you see a pedestrian at a zebra or pelican crossing carrying a white stick, do you think . . .   Tick the appropriate box

☐ 1 He/she has difficulty walking?

 2 He/she is visually handicapped?

Q11 The traffic lights at a pelican crossing have the same meaning as ordinary traffic lights, but they do not have a red and amber phase.

1 What do they show instead of the red and amber phase?   Answer _Flashing Amber_

2 What does the light mean?   Answer _You must give way to Pedestrians_

Q12 What sound is usually heard at a pelican crossing when the green man is shown to pedestrians?

Answer _A Bleeping tone_

ANSWERS ON PAGE 119

A level crossing is where the road crosses at a railway line. It is potentially dangerous and should be approached with caution.

Q13 Match each traffic sign below with its correct meaning.

A B C D

KEEP CROSSING CLEAR

1 Level crossing without gates or barriers — B

2 Level crossing with lights — C

3 Level crossing with gates or barriers — A

4 Level crossing without lights — D

Q14 If you break down on a level crossing, should you . . .

1 Tell your passengers to wait in the vehicle while you go to get help?

2 Get everybody out and clear of the crossing? ✓

3 Telephone the police?

4 Telephone the signal operator? ✓

5 If there is time, push your car clear of the crossing? ✓

Tick the appropriate boxes

ANSWERS ON PAGE 119

One-way systems are where all traffic flows in the same direction.

Q1 Which of these signs means one-way traffic?

A

B

Answer

Q2 Are these statements about one-way systems true or false?

TRUE   FALSE

 1 In one-way streets traffic can pass me on both sides.

 2 Roundabouts are one-way systems.

3 For normal driving I should stay on the left.

4 I should look out for road markings and get in lane early.

Tick the appropriate boxes

ANSWERS ON PAGE 119

As a rule, the more paint on the road, the more important the message.

Q3 Road markings are divided into three categories.

Fill in the missing words

1 Those which give i_N_F_O_R_M_A_I_I_O_N_.

ANSWERS ON PAGE 119

2 Those which give w_A_R_N_I_N_G_S_.

3 Those which give o_R_D_E_R_S_.

Q4 There are two main advantages which road markings have over other traffic signs. Name them

1 They can be seen when other signs are hidde

2 They give a Continuous message

Q5 What do these lines across the road mean?

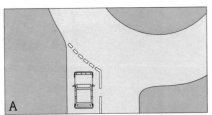

1 Stop and give way

2 Give priority to traffic coming from the immediate right.

3 Give way to traffic coming from the right.

Answer

1 Give way to traffic on the major road.

2 Stop at the line and give way to traffic on the major road.

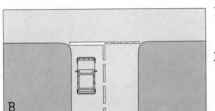

Answer

Q6 What do these lines painted along the road mean?

[✓] 1 I must not park or wait on the carriageway.

[ ] 2 I can park between 7pm and 7am

[ ] 3 I must not overtake.

[✓] 4 I must not cross the white line except to turn right or in circumstances beyond my control.

More than one answer may be correct. Tick any boxes you think are appropriate.

Q7 What is the purpose of these hatched markings (chevrons)?

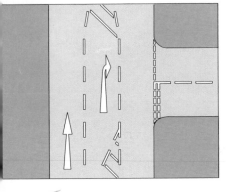

Answer _They are used to separate potentially dangerous streams of traffic_

ANSWERS ON PAGE 119

Q8 What does it mean if the chevrons are edged with a solid white line?

Answer _You must not enter this_

The shape and colour of a sign will help you understand what it means.

Q9   Look at the signs below and say whether each gives an order,
a warning or information.

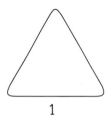

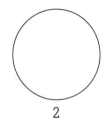

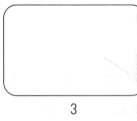

1   2   3

Answer _Warning_   Answer _order_   Answer _information_

Q10   1 A circular sign with a blue background tells you what you
m_ust_ do.

2 A circular sign with a red border tells you what you m_ust_ n_ot_ do.

Complete the sentences

Q11   What do these signs mean?

1   Answer _Give way to traffic on a Major
              Road._

2   Answer _Stop and give way_

ANSWERS ON PAGE 120

Q12   Some junctions have a stop sign, others have a give way sign.

Complete the sentence

A stop sign is usually placed at a junction where v i s i o n

s l i m i t e d .

Q13   Information signs are colour-coded.

Match each of the following five types of sign to its colouring

D Motorway signs   A White letters on a brown background

E Primary routes   B Black letters on a white background

B Other routes   C Black letters on a white background with
                                a blue border

C Local places   D White letters on a blue background with
                                a white border

A Tourists signs   E White letters on a green background,
                                yellow route numbers with a white border

ANSWERS ON PAGE 120

Q1   Good observation is vital in today's busy traffic.

Complete the sentence

When using my mirrors I should try to make a mental note of the s _peed_,

b_ehaviour_ and i_ntentions_ of the driver behind.

Q2   Driving in built-up areas is potentially dangerous.

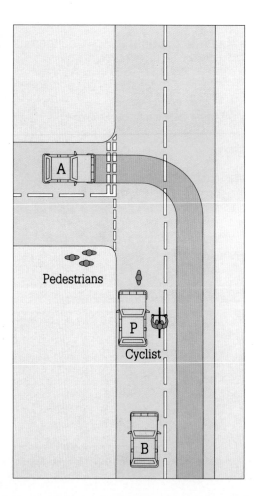

Look at this diagram

1 What action should the driver of

   car A take? List four options

   A _Observation_

   B _Pedestrians_

   C _Cyclist_

   D _Car B_

2 What action should the driver of

   car B take? List four options

   A _Slow down_

   B _Pedestrians_

   C _Pass Car_

   D _Car A_

ANSWERS ON PAGE 120

Q3 Motor cyclists are often less visible than other road users.

Complete this well-known phrase

Think once, think twice, think b_ _K _E_.

Q4 When you observe traffic following too close behind you,

would you . . .   Tick the appropriate box

☐ 1 Speed up to create a bigger gap?

☐ 2 Dab your brake lights to warn the following driver?

☑ 3 Keep to a safe speed, and keep checking the behaviour and intentions
of the following driver?

Q5 Some hazards are potential, others are actual and there all the

time, such as a bend in the road.

A   Name five more actual hazards   B   Name five potential hazards, such

1 _Junctions_   as a dog off its lead

2 _hump back bridge_   1 _Children playing_

3 _Concelled entrance_   2 _horses_

4 _Dead ground_   3 _Pedestrians_

5 _Narrow lanes_   4 _Eddrley / young_

                                 5 _Cyclist / other road users_

ANSWERS ON PAGE 120

One of the features of driving on the open road is taking bends properly.

Q6 As a rule you should be travelling at the correct s_peed_,
using the correct g_ear_, and be in the correct p_osition_ .

Q7 Should you brake . . .

☑ 1 Before you enter the bend?   ☐ 2 As you enter the bend?

☐ 3 While negotiating the bend?

Tick the appropriate box

Q8 Which way does force push a car on a bend?

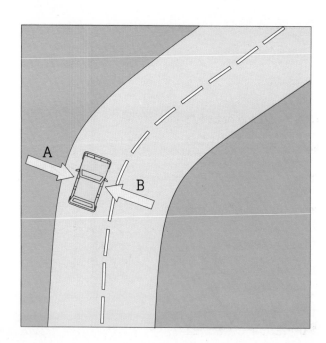

A Inwards

or

B Outwards

Answer

Q9 What happens to the weight of the car when you use the brakes?

A It is thrown forwards B It remains even C It is thrown back Answer

Q10 When you approach a bend, what position should you be in?

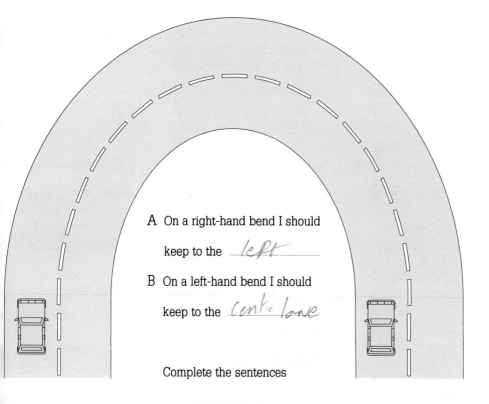

A On a right-hand bend I should keep to the _left_

B On a left-hand bend I should keep to the _centre lane_

Complete the sentences

ANSWERS ON PAGE 121

Overtaking is a potentially dangerous manoeuvre.

Q11 Before overtaking, consider whether it is really

ne_cessary_.   Fill in the missing word

Always use the safety routine when overtaking.

Put these actions into their correct sequence by putting the numbers 1 to 7, as seen in

the diagram, in the boxes

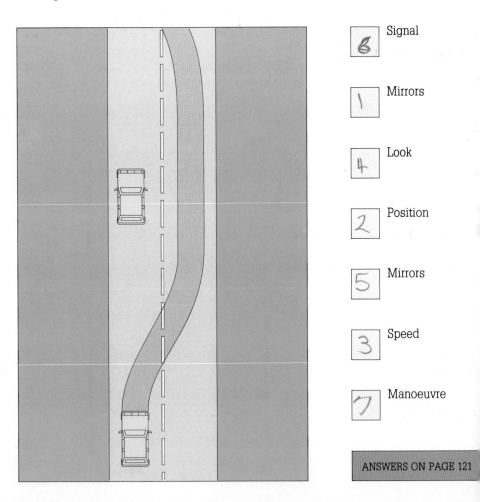

6	Signal
1	Mirrors
4	Look
2	Position
5	Mirrors
3	Speed
7	Manoeuvre

ANSWERS ON PAGE 121

Q12 What is the minimum amount of clearance you should give a

cyclist or motor cyclist?

Answer _about the width of a small car_

Q13 There are four situations in which you may, with caution,

overtake on the left-hand side of the car in front.   Name them

1 _The one vehicle in front is signalling and_
posistion to turn Right

2 _You are using the correct lane to turn_
left at a junction

3 _Traffic is moving slowly in queue, and the_
Traffic on the right is moving more slowly

4 _You are in a One way Street_

Q14 List four places where it would be dangerous to overtake.

1 _On approach to a Junction_

2 _The brow of a hill_

3 _The approach to a bend_

4 _Where there is dead ground_

ANSWERS ON PAGE 121

Q15 Dual carriageways can appear similar to motorways, but there are important differences.

Which of the following statements apply to dual carriageways?

Tick the relevant boxes

☐ 1 Reflective studs are not used.

☑ 2 Cyclists are allowed.

ANSWERS ON PAGE 121

☐ 3 The speed limit is always 60 mph.

☐ 4 You cannot turn right to enter or leave a dual carriageway.

☐ 5 Milk floats and slow moving farm vehicles are prohibited.

Q16 When turning right from a minor road on to a dual carriageway, where would you wait . . .

A When there is a wide central reserve?

Answer *You would cross over the first carriage way then wait for gap in central reservations.*

B When the central reserve is too narrow for your car?

Answer *You wait untill a gap in traffic,*

Q17 When travelling at 70 mph on a dual carriageway, which lane would you use?

Answer *Speed limit applying to all lanes*

Q18 What do these signs mean?

A

B

C

Answer *Dual carriageway ends*　　Answer *Road narrows on both sides*　　Answer *Two way traffic ahead*

Q19 Which of the above signs would you expect to see on a dual carriageway?

Answer *A and C*

ANSWERS ON PAGE 121

Q1 Cars fitted with automatic transmission select the gear depending on the road speed and the load on the engine.

They therefore have no c_Lutch_ pedal.

Fill in the missing word

Q2 The advantages of an automatic car are . . .

1 _Driving is easier_

2 _Thre is more time to Conentrate on road_

Q3 The gear selector has the same function as a manual selector, but what function do each of the following have?

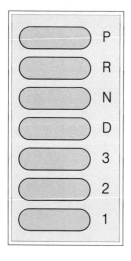

Park _lacks transmission_

Reverse _enable car to go backwa_

Neutral _Same function asmanual._

Drive _driving car forward_

3rd ⎫

2nd ⎬ _have the same_

1st ⎭ _function as Manad gears_

ANSWERS ON PAGE 122

Q4 Automatic cars have a device called a kickdown. Is its function . . .

1 To select a higher gear?

2 To select a lower gear manually?

3 To provide quick acceleration when needed? Tick the relevant box

Q5 When driving an automatic car, would you select a lower gear . . .

YES NO

1 To control speed when going down a steep hill?

2 To slow the car down in normal driving?

3 When going uphill?

4 To overtake, in certain circumstances?

5 When manoeuvring?

6 Before stopping?

ANSWERS ON PAGE 122

Tick the appropriate boxes

Q6 An automatic car has two foot pedals, the foot-brake and the accelerator.

For normal driving, which foot would you use . . .

1 For the brake?   Answer *Right*

2 For the accelerator?   Answer *Right*

Q7 When driving along using one foot to control both pedals is preferable to using both the left and the right foot. Why?

Answer *Stops you using brake and accellerator at the same time*

Q8 Some cars with automatic transmission have a tendency to 'creep'.

Which gears allow the car to creep?

Answer *Drive, Reverse all gears'*

ANSWERS ON PAGE 122

Q9 When driving an automatic car, would you use the handbrake . . .

☑ 1 More than in a manual car?

☐ 2 The same?

☐ 3 Less?   Tick the relevant box

Q10 In which position should the gear selector be when you are starting the engine?

Answer ___Parking___ or ___Neutral___

Q11 As you approach a bend, an automatic car will sometimes change up because there is less pressure on the accelerator.

What should you do to prevent this happening?

☑ 1 Slow down before the bend and accelerate gently as you turn.

☐ 2 Brake as you go round the bend.

☐ 3 Brake and accelerate at the same time.

Tick the relevant box

ANSWERS ON PAGE 122

Q1 There are many myths and legends surrounding the driving test.

Are the following statements true or false? Tick the relevant boxe

TRUE FALSE

☐ ☑ 1 The driving test is designed to see whether I can drive around a test route without making any mistakes.

☑ ☐ 2 The driving test is designed to see whether I can drive safely under various traffic conditions.

☑ ☑ 3 I do not need to know any highway code.

☐ ☑ 4 The examiner has a set allocation of passes each week.

☑ ☐ 5 My knowledge of other motoring matters, for example the cause of skids, may be tested.

Q2 The length of the normal driving test is approximately . . .

☐ 1 60 minutes

☐ 2 90 minutes

☑ 3 35 minutes

Tick the correct box

ANSWERS ON PAGE 122

Q3 If, during the test, you do not understand what the examiner says to you, you should take a guess because you must not talk to him or her.

TRUE   FALSE

Is this statement true or false?   Tick the correct box

Q4 You may have heard people say that it is easier to pass the driving test in certain parts of the country.

YES   NO

Do you agree with this statement?   Tick the correct box

Q5 If you fail your test, you can take it again.

Which of the following statements is correct?   Tick the relevant box(es)

1 If you fail the test, you can apply straight away for another appointment.

2 If you fail the test you have to wait a month before you can apply for another appointment

3 You can re-take your test, subject to appointment availability, any time.

4 You have to wait a month before you can re-take the test.

ANSWERS ON PAGES 122-23

Q6 Before the practical part of your test, the examiner will test your eyesight.   This is done by asking you to read a number plate at a distance of . .

☐ 1 30.5 metres (100 feet)

☑ 2 20.5 metres (67 feet)

☐ 3 40.5 metres (133 feet)   Tick the correct box

Q7 What will happen if you fail your eyesight test?

Answer *The test will not procced*

Q8 It is essential that you take your p*rovisonal*
li*cence* to the test centre. This document must be
s*igned* in ink.   Complete the sentences

Q9 The examiner will expect you to drive without making any mistakes.   Do you think this statement is true or false?

TRUE   FALSE
☐   ☑
 Tick the correct box

ANSWERS ON PAGE 123

Q10 Is this statement about what you will be asked to do during the test true or false?

I will be asked to perform four set exercises:

1 The emergency stop

2 The turn in the road

3 Reversing into a side road on the right or left

4 Reverse parallel parking

TRUE   FALSE

☐ ☑ Tick the relevant box

Q11 When reversing, are you allowed to undo your seat belt?

TRUE   FALSE

☑ ☐ Tick the correct box

Q12 If you fail your test, what will the examiner do?

1 give you verbal explanation

2 might a formal Failure Sheet

ANSWERS ON PAGE 123

Q13 When you have passed your driving test, what are

you entitled to do?

1 _Drive unsupervised_

2 _Drive on Motorway_

3 _Drive without L plates_

Q14 I have within the last month passed my test.

Can I supervise a learner driver?

YES   NO

☐   ☑ Tick the correct box

Q15 When you pass your test, where should you send

your pass certificate?

Answer _DVLC SWANSEA_

Q16 While you are waiting for your full licence to be sent to you,

can you drive legally?

YES   NO

☑   ☐ Tick the correct box

ANSWERS ON PAGE 123

Q17 It is recommended that you take further tuition once you have passed your test, especially on motorway driving.

As a learner driver you will not have experienced the special r_ules_ · that apply on the motorway and the h_igh_ s_peed_ of the other traffic.

Complete the sentence

Q18 While taking your driving test, you should drive . . .

[ ] 1 Especially carefully, keeping about 5 mph below the speed limit

[✓] 2 As you would normally drive with your instructor

[ ] 3 With confidence, keeping at or just over the speed limit, to show that you can really drive.

Tick the relevant box

Q19 Can you take a driving test if you are deaf?

YES   FALSE
[✓]   [ ]   Tick the correct box

ANSWERS ON PAGE 123

The driving test ensures that all drivers reach a minimum standard.

Q1 Do you think that learning to drive ends with passing the test?

YES   NO

☐ ☑ Tick the correct box

Q2 What knowledge and skills are not necessarily assessed in the present driving test?   List three

1 _Bad weather driving_
2 _Night time driving_
3 _Motorway driving_

Q3 Which of these statements do you think best describes advanced driving?   Tick the relevant box

☐ 1 Advanced driving is learning to handle your car to its maximum performance.

☑ 2 Advanced driving is learning to drive defensively with courtesy and consideration to others.

☐ 3 Advanced driving is learning to drive fast.

ANSWERS ON PAGE 123

Q4   Some people have difficulty in driving at night.

Who would you expect, in general, to experience most difficulties?

Tick the relevant box

[✓] 1 Older people

[ ] 2 Younger people

Q5   Once you have passed your driving test, your licence is usually

valid until you reach *70* years of age.   Complete the sentence

Q6   There are particular circumstances under which you are required

to take a driving test again.   Name them

Answer   *Serious driving offences and*
illness after 10 years.

ANSWERS ON PAGE 124

Q7 Motorways are designed to enable traffic to travel faster in greater safety. Compared to other roads, are they statistically . . .

☑ 1 Safer   ☐ 2 Less safe?   ☐ 3 No different

Tick the relevant box

Q8 Are the following groups allowed on the motorway?

YES   NO

☐ ☑ 1 Provisional licence holders

☑ ☐ 2 Motor cycles over 50cc

☐ ☑ 3 Pedestrians

☑ ☐ 4 HGV learner drivers

☑ ☐ 5 Newly qualified drivers with less than three months' experience

☑ ☐ 6 Motor cycles under 125cc

☐ ☑ 7 Cyclists

Tick the relevant boxes

ANSWERS ON PAGE 124

Q9 There are some routine checks you should carry out on your car before driving on the motorway. Name four of them

1 _oil_ 2 _Water_

3 _Fuel_ 4 _Tyre pressure_

Q10 On the motorway, if something falls from either your own or another vehicle, should you . . .

☐ 1 Flash your headlights to inform other drivers?

☐ 2 Pull over, put your hazard warning lights on and quickly run on to the motorway to collect the object?

☑ 3 Pull over on to the hard shoulder, use the emergency telephone to call the police?

☐ 4 Flag another motorist down to get help?

Q11 Which colour do you associate with motorway signs?

☐ 1 Black lettering on a white background

☐ 2 White lettering on a green background

☑ 3 White lettering on a blue background

ANSWERS ON PAGE 124

Tick the correct box

Q12 At night or in poor weather conditions, your headlights will pick out reflective studs. Match the colour of the studs to their function by placing the appropriate letter in the box.

Box	Colour		Function
B	Amber	A	Marks the edge of the hard shoulder
A	Red	B	Marks the edge of the central reservation
D	Green	C	Marks lane lines
C	White	D	Marks exits and entrances

Q13 Do the broken lines at the end of the acceleration lane mean . . .

[ ] 1 The edge of the carriageway?

[ ] 2 Other traffic should let you in?

[✓] 3 Give way to traffic already on the carriageway?   Tick the correct box

Q14 If you see congestion ahead, is it legal to use your hazard warning lights to warn drivers behind you?   Tick the correct box

YES [✓]   NO [ ]

ANSWERS ON PAGE 124

Q15   What is the most common cause of accidents on motorways?

☐ 1 Vehicles breaking down   ☐ 2 Drivers falling asleep

☑ 3 Drivers travelling too fast, too   ☐ 4 Fog   Tick the correct box
      close to the vehicle in front

Q16   Are the following statements true or false?

can use the hard shoulder . . .

TRUE	FALSE		
☐	☑	1 To take a short break	
☐	☑	2 To stop and read a map	
☐	☑	3 To allow the children to stretch their legs	
☑	☐	4 To pull over in an emergency	
☐	☑	5 To answer a phone call	Tick the relevant box

ANSWERS ON PAGE 124

Q17   In normal driving on the motorway, you should overtake . . .

☑ 1 On the right   ☐ 2 On the left

☐ 3 On either side

Tick the correct box

Driving at night can cause problems.

ANSWERS ON PAGE 124

Q18   Which of these statements do you think is correct?

Tick the relevant box

☐ 1 Street lighting and my car's headlights mean that I can see just as well as in the daylight. Therefore driving at night is just like driving in the daylight.

☑ 2 At night I have to rely on my car's headlights and any additional lighting. Therefore I cannot see as far or drive as fast as in the daylight.

Q19   At dusk and dawn what action should you take to compensate for driving a dark coloured car?

Answer _Switch on earler / switch off later_

Q20   When driving after dark in a built-up area, should you use . . .

☑ 1 Dipped headlights?

☐ 2 Side or dim-dipped lights?   Tick the correct box

Q21   The Highway Code says you should not use your horn in a built-up area between 11.30pm and 7am.

What is the exception to that rule?

Answer _If you are stationary to aviod danger from moving vehicle_

Q22 The diagram below illustrates two vehicles parked at night on a two-way road.   Which one is parked correctly?

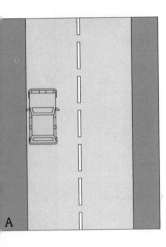

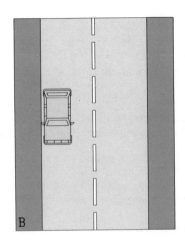

A   B   Answer [*A*]

Q23 Certain groups of road users are particularly vulnerable at night.

Name two of them

1 _Pedestrians_   2 _cyclist_.

Q24 Under what circumstances would you use dipped headlights during the day?   Answer _In poor weather_

S_ee_ and b_e_ s_een_.   Complete the sentence

Q25 When you are waiting at a junction after dark, your brake lights

might d_azzle_ the driver behind. It is better to

use your h_andbrake_. Complete the sentences

ANSWERS ON PAGES 125-26

Certain weather conditions can create hazardous driving conditions in the summer as well as in the winter.

Q26 Which of the following causes greatest danger to drivers?

Tick the correct box

☐ 1 Snow

☐ 2 Ice

☐ 3 Heavy rain

☑ 4 Not being able to see properly

Q27 In wet weather conditions your tyres can lose their grip.
You should allow at least d_double_ the distance between you and the car in front that you allow on a dry road.

Fill in the missing word

Q28 In very wet conditions there is a danger of a build-up of water between your tyres and the road.
This is called a _quaplaning_.

Fill in the missing word

ANSWERS ON PAGE 125

Q29 How should you deal with floods?

Tick the correct box

☐ 1 Drive through as fast as possible to avoid stopping

☑ 2 Drive through slowly in 1st gear, slipping the clutch to keep the engine speed high

☐ 3 Drive through in the highest gear possible, slipping the clutch to keep the engine speed high

Q30 Will less tread on your tyres . . .   Tick the correct box

☑ 1 Increase your braking distance?   ☐ 2 Decrease your braking distance?

Q31 When the tyres lose contact with the road, the steering will feel

very light.   Complete the sentence

Q32 After you have driven through a flood, should you check . . .

☐ 1 Your speedometer?   ☑ 2 Your brakes?

☐ 3 Your oil?

ANSWERS ON PAGE 125

Tick the correct box

Q33 There are certain key precautions you should take when driving in fog. Complete the following sentences

1 S_low_ d_own_.

2 Ensure you are able to s_top_ within the distance you can see to be clear.

3 Use your w_INDSCREEN_ w_IPERS_

4 Use your d_emister_ and your h_eated_ r_ear_ w_indscreen_

Q34 Under what circumstances should you use your rear fog lights?

When visibility is less than ___100___ metres/yards

Fill in the correct number

Q35 When you are following another vehicle in fog, should you . . .

☐ 1 Follow closely behind because it will help you see where you are going?

☑ 2 Leave plenty of room between you and the vehicle in front?

Tick the correct box

ANSWERS ON PAGE 125

Q36 When you are following another vehicle in fog, should you use . . .

☐ 1 Main beam headlights?　　　☑ 2 Dipped headlights?

Tick the correct box

Q37 Extra precautions are needed when dealing with a junction in fog.

Complete the following sentences

1 Open your w_I_N_D_O_W_S and switch off your a_u_d_i_o_ s_y_s_t_e_m.
L_I_S_t_e_N for other vehicles.

2 Signal e_a_r_L_y .

3 Use your b_r_a_k_e_s. The light will a_L_e_r_t following vehicles.

4 Use your h_o_R_N if you think it will w_a_r_N other road users.

Q38 Is the following statement about anti-lock brakes true or false?

TRUE　FALSE
☐　　☑ Anti-lock brakes will stop me skidding when driving on snow or ice.

Tick the correct box

ANSWERS ON PAGE 125

Q39 When driving in snow or ice you should gently test your

b r a k e s from time to time.   Fill in the missing word

Q40 In order to slow down when driving on snow or ice you should . . .

1 Use your brakes g e n t l y.

2 Get into a l o w e r g e a r earlier than normal.

3 Allow your speed to d r o p and use your b r a k e s gently and early.

Fill in the missing words

Q41 On snow or ice braking distances can increase by . . .

1 10 times ✓

2 5 times

ANSWERS ON PAGE 125

3 20 times

4 15 times   Tick the correct box

Q42 When going downhill in snow, what would you do to help you slow down?

Answer If possible, control your speed before reaching the hill select low gear early

Q43

When cornering in snow or ice, what should you avoid doing?

Answer _using your brakes_

Q44

How can you reduce the risk of wheel spin?

Answer _Aviod harsh accelleration_

Q45

Three important factors cause a skid.   Name them

1 _The driver_

2 _The veduch_

3 _The road Condutions._

Q46

Some everyday driving actions, especially in poor weather, can increase the risk of skidding.

Fill in the missing words

1 S_lowing_ down.

2 S_peeding_ up.

3 T_urning_ corners.

4 Driving u_phill_ and d_ownhill_.

ANSWERS ON PAGE 125

All vehicles need routine attention and maintenance to keep them in good working order. Neglecting maintenance can be costly and dangerous.

Q1 With which of these statements do you agree? Tick the relevant box

[✓] 1 Allowing the fuel gauge to drop too low is bad for the engine.

[ ] 2 In modern cars the fuel level makes little difference.

Q2 What do you put into the engine to lubricate the moving parts?

Answer Oil

Q3 How frequently should you check your oil level? Tick the correct box

[ ] 1 Once a month

[ ] 2 Once a year

ANSWERS ON PAGE 126

[✓] 3 Every time you fill up with fuel

Q4 The engine is often cooled by a mixture of water and a_Nti_-f_reeze_. Some engines are a_ir_ cooled.

Complete the sentences

Q5   How frequently should you test your brakes?   Tick the correct box

[✓] 1 Daily   [ ] 2 Monthly   [ ] 3 Weekly

[ ] 4 When I use them

ANSWERS ON PAGE 126

Q6   Incorrectly adjusted headlamps can cause dazzle to other road users.   Complete the sentence

Q7   All headlamps, indicators and brake lights should be kept in good working order.

It is also important that they are kept clean   Fill in the missing word

Q8   Tyres should be checked for uneven wear and tyre walls for bulges and cuts.   Complete the sentence

Q9   The legal requirement for tread depth is not less than . . .

[ ] 1 1.4mm   [✓] 2 1.6mm   [ ] 3 2mm   Tick the correct box

Q10   What should you do if your brakes feel slack or spongy?

Answer Get them checked quickly

It is important that all road users know what to do in the case of breakdown, accident or emergency. A lack of knowledge could put lives at risk.

Q11 Vehicle breakdowns could result from . . .

1 N_eglect_ of the vehicle

2 Lack of r_outine_ c_hecks_

3 Little or no p_reventative_ maintenance

4 A_buse_ of the vehicle

Fill in the missing words

Q12 It is advisable to carry a warning triangle.

1　On a straight road how far back should it be placed?

[✓] 50m/yds　　　[ ] 200m/yds　　　[ ] 150m/yds

Tick the correct box

2　On a dual carriageway, how far back should it be placed? At least . . .

[ ] 200m/yds　　　[✓] 150m/yds　　　[ ] 400m/yds

Tick the correct box

ANSWERS ON PAGE 126

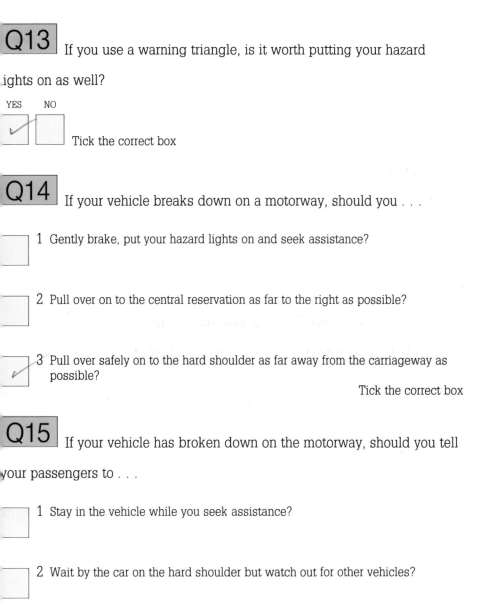

Q13 If you use a warning triangle, is it worth putting your hazard lights on as well?

YES   NO

Tick the correct box

Q14 If your vehicle breaks down on a motorway, should you . . .

1 Gently brake, put your hazard lights on and seek assistance?

2 Pull over on to the central reservation as far to the right as possible?

3 Pull over safely on to the hard shoulder as far away from the carriageway as possible?

Tick the correct box

Q15 If your vehicle has broken down on the motorway, should you tell your passengers to . . .

1 Stay in the vehicle while you seek assistance?

2 Wait by the car on the hard shoulder but watch out for other vehicles?

3 Get out of the vehicle and wait on the embankment away from the hard shoulder?

ANSWERS ON PAGE 126

Tick the correct box

Q16 The marker posts at the side of all
motorways have a picture of a telephone handset.
How can you tell which way to walk to reach
the nearest telephone?

Answer _Under handset points to the nearest telephone_

Q17 When you use the emergency telephone on a motorway, what
will the operator ask you?

1 _The emergency number painted on box_
2 _Vehicle details (make, model, registration_
3 _Membership details_
4 _Details of fault_

Q18 Disabled drivers cannot easily get to an emergency telephone.
How can they summon help?

1 _By display help pendant_
2 _Using mobile phone_

ANSWERS ON PAGE 126

Q19 If you break down when travelling alone, there are three things

you are advised NOT to do.   Complete these sentences

Do not ask p_assing_ m_otorist_ for help.

Do not accept help from anyone you d_o_ n_ot_ k_now_ (except the

emergency services or a breakdown service).

Do not l_eave_ your vehicle l_onger_ than necessary.

Q20 If I am first or one of the first to arrive at the scene of

an accident, should I . . .

TRUE   FALSE

1 Always move injured people away from vehicles? [FALSE ✓]

2 Tell the ambulance personnel or paramedics what I think is wrong with those injured? [FALSE ✓]

3 Give casualties something warm to drink? [FALSE ✓]

4 Switch off hazard warning lights? [FALSE ✓]

5 Switch off vehicle engines? [TRUE ✓]

6 Inform the police of the accident? [TRUE ✓]

ANSWERS ON PAGE 126

Tick the appropriate boxes

Q21 If you are involved in an accident, what MUST you do?

Answer _Stop_

Q22 If you are involved in an accident and nobody is injured, do you have to call the police?

YES   NO

☐ ☐ Tick the correct box

Q23 What information do you need to exchange if you are involved in an accident?

1 _The other drivers name and address + tele_
2 _The registration number of all vechile involed_
3 _The make of the other car_
4 _The other driver insurance details_
5 _If he drove not the owner, Owner deta_

ANSWERS ON PAGES 126-27

Q24 If you thought you had a fire in your car's engine, what action would you take?

1 _Pull up quickly_

2 _get passenger out_

3 _Call assistance ._

Q25 There are three items of emergency equipment it is wise to carry in your car.

1 F_irst_ a_id_ kit.

2 F_ire_ e_xtinguisher._

3 W_arning_ t_riangle_.

Fill in the missing words

Q26 When you rejoin a motorway from the hard shoulder, should you . . .

☐ 1 Signal right and join when there is a safe gap?

☐ 2 Keep your hazard lights on and drive down the hard shoulder until there is a safe gap?

☑ 3 Use the hard shoulder to build up speed and join the carriageway when safe?

Tick the correct box

ANSWERS ON PAGE 127

Q27 Fuel combustion causes waste products.

One of these is a gas called c_o__r__b__e__n___ d_i__o__x__i__d__e___. This is a major

cause of the g_r__e__e__n___ h__o__u__s__e___ effect.

Complete the sentences

Q28 How much does transport contribute to the production of carbon

dioxide in this country (expressed as a percentage of the total production)?

☐ 1 10 per cent   ☐ 2 25 per cent

☐ 3 50 per cent   ☑ 4 20 per cent

Q29 The MOT test checks the roadworthiness of a vehicle.

Does it include an exhaust emission test?

YES   NO
☑   ☐   Tick the correct box

Q30 A catalytic convertor stops the emission of carbon dioxide.

TRUE   FALSE
☐   ☑

Tick the correct box

ANSWERS ON PAGE 127

Q31 Which is more environmentally friendly?

☐ 1 A petrol engine with a catalytic convertor

☒ 2 A diesel engine

Tick the correct box

ANSWERS ON PAGE 127

Q32 Which uses up more fuel?

☐ 1 A car travelling at 50mph

☒ 2 A car travelling at 70mph

Tick the correct box

Q33 Which is environmentally less harmful?

☐ 1 Leaded fuel

☒ 2 Unleaded fuel

Tick the correct box

Q34 There are some measures car drivers can take to help reduce damage to the environment.

List five

1 _Make sure vechile is in good condition Surviced Regular_

2 _Make sure tyres are Correct inflated and in flight waste fuel_

3 _Push Choke in asAP_

4 _Avoid hard braking_

5 _Buy Fuel efficant vechile_

Before buying a used car it is best to decide what you want the car for and how much you can afford.

Q35 There are three main sources of supply for used vehicles. You can buy from a d_ealer_ , at an a_uction_ or p_rivately_. Complete the sentence

Q36 When reading a glowing description of a used car, what should you first consider?

Answer _Why it is being sold_

Q37 Are these statements about buying a used car through a dealer or at an auction true or false?

TRUE   FALSE

1 It is often cheaper to buy a car at an auction than through a dealer.

2 I have the same legal rights when I buy at an auction as when I buy from a dealer.

3 I should always read the terms and conditions of trade before I buy a car at an auction.

4 The best way to select a used car dealer is by recommendation.

Tick the appropriate boxes

ANSWERS ON PAGE 127

Q38

Cars bought through a dealer often have a warranty.

What should you check?

What is Covered

The lenght of agreement

Q39

When you test drive a vehicle, you should make sure that it is _axed_, has a current M_OT_ certificate (if applicable) and that all _nsurance_ requirements are complied with.

Complete the sentence

Q40

There are some important items that you should check on before you buy a used car.   List three

1 _Milage_

2 _If car has been involed in accident_

3 _Number of owners_

Q41

Do you think the following statement is true or false?

It is advisable to have my vehicle examined by a competent and unbiased expert before I buy.

TRUE   FALSE

☑ ☐

Tick the correct box

ANSWERS ON PAGES 127-28

Particular difficulties are encountered when towing a caravan or trailer. There are some very good courses which will help master the skills required.

Q1 People can underestimate the length of the total combination of car and caravan or trailer. Is the overall length usually . .

☑ 1 Twice the length of a normal car?

☐ 2 Three times the length of a normal car?

Tick the correct box

Q2 What additional fixtures should you attach to your car to help you see more clearly?

Answer _Extua towirg Mirror give you moe view_

Q3 When towing you will need more distance than normal to overtake. Is it . . .

☐ 1 Twice the normal distance?

☑ 2 Three times the normal distance?

☐ 3 Four times the normal distance?

Tick the correct box

Q4 A device called a s_tabliges_ will make the combination safer to handle.

Fill in the missing word

ANSWERS ON PAGE 128

Q5 The stability of the caravan will depend on how you load it.

should heavy items be loaded . . .   Tick correct box

☐ 1 At the front?   ☐ 2 At the rear?   ☑ 3 Over the axle(s)?

Q6 There are special restrictions for vehicles which are towing.

A What is the speed limit on a dual carriageway?   Tick the correct box

☐ 1 50mph   ☑ 2 60mph   ☐ 3 70mph

B What is the speed limit on a single carriageway?   Tick the correct box

☐ 1 40mph   ☑ 2 50mph   ☐ 3 60mph

Q7 There are some important checks you should make before

starting off.   List four

1 _loading_   2 _lights_

3 _hitched_   4 _tyres_

Q8 If you decide to stop to take a break, before allowing anyone to

enter the caravan you should lower the j_ockey_ w_heel_ and

corner s_tabilisers_.

Fill in the missing words

ANSWERS ON PAGE 128

Many people now take their car abroad or hire a vehicle when on holiday.

Q9 Motoring organisations, such as The Automobile Association, can help you plan and organise your trip.

The AA can provide advice on travel and v_e_c_h_i_l_c_e insurance. They will also help you to organise the d_oc_u_m_e_n_t_s that you will need.

Fill in the missing words

Q10 Before venturing on to the Continent, you should always . . .

1 Plan the r_o_u_t_e you wish to take.

2 Know the local m_o_t_a_r_i_n_g r_e_g_u_l_a_t_i_o_n_s.

Complete the sentences

Q11 It is essential that your vehicle should be checked thoroughly.

List four of the routine checks you should make.

1 Tyres

2 Tool kit

3 lamps

4 kit deflectas

ANSWERS ON PAGE 128

Q12 In most European countries you are advised to carry your

living __ *licence* on you.   Complete the sentence

Q13 What do the letters I D P stand for?

Answer *International Driving Permit*

Q14 Where might you need an I D P ?

Answer *Some Non EC countries*

Q15 In most European countries what age do you have to be to drive?

☐ 1 21   ☑ 2 18   ☐ 3 16

Tick the correct box

Q16 Some European countries can require you to carry additional emergency equipment.

List four of the items you are recommended to carry

1 *Spare lamps / bulbs*

2 *Warning triangle*

3 *First Aid kit*

4 *Fire Extinguisher*

Emergency windscreen

ANSWERS ON PAGE 128

Part 1

Introduction to learning to drive
Questions on pages 10-11

A1

A current, signed, full or provisional licence for the category of vehicle that you are driving

A2

examinations register

A3

21 years old
three years

A4

To the front and rear. It is important not to place them in windows where they could restrict good vision.

A5

True

A6

You should have answered No to all the questions.

A7

Yes. This should be the ambition of every driver.

Adjusting your driving position
Question on page 12

A8

1   handbrake
2   doors
3   seat
4   head restraint
5   mirrors
6   seat belt

Introduction to vehicle controls
Questions on pages 13 - 14

A9

The handbrake	E
The driving mirrors	D
The gear lever	F
The clutch	G
The steering wheel	A
The foot-brake	B
The accelerator	C
gas	

A10

The foot-brake	R
The clutch	L
The accelerator	R

A11

1   False. You will need one hand to change gear or use other controls.
2   True
3   False. The best position is a quarter to three or ten to two.
4   False. It is safest to feed the wheel through your hands.
5   True
6   True

A12

The direction indicators	H
Dipped beam	A
Main beam	D
Rear fog lamp	C
Horn	H
Hazard lights	N

Part 2

Moving off
Questions on page 15

A1

A	1
B	5
C	3
D	2
E	6
F	4
G	7
H	8
I	9

art 17   **Answers**   113

topping
(ormally)
uestion on page 16

A2

1
3
2
4
6
5
8
7
9

ear changing
uestions on pages 17 - 20

A3

st gear

A4

h, or 4th if the car has
4-speed gear box

A5

sually 2nd gear, but 1st if
u need to go very slowly
3rd if the corner is
weeping and you can
ke it safely at a higher
eed

A6

gine
hicle
und
hen

A7

A   1
B   3
C   2
D   4
E   5

A8

1   False.   This will
cause the engine to
labour.
2   False.   It is good
practice to use the brakes
to slow the car down.
Using the transmission
causes wear and tear
which can be very costly.
Also, the brakes are more
effective.
3   False
4   True
5   False.   It is good
practice to miss out the
unwanted gears and
select the gear most
appropriate to your
road speed.

A9

1

A10

1   Don't
2   Don't
3   Do
4   Do
5   Do
6   Don't
7   Don't
8   Don't

Steering
Questions on
page 21

A11

A,   except in a
few cars fitted with
four-wheel steering
(in which case all
four wheels
will move)

A12

A

A13

B

A14

B

Road positioning
Questions on
page 22

A15

C,   well to the left
but not too close
to the kerb

A16

B.   Avoid swerving
in and out.
It is unnecessary
and confuses
other drivers.

Clutch control
Questions on page 23

A17

A

A18

biting

A19

1 Yes
2 Yes
3 No
4 Yes
5 No

Part 3

Junctions
Questions on pages 24 - 27

A1

two or more roads

A2

A T Junction
B Y Junction
C Roundabout
D Staggered crossroads
E Crossroads

A3

1 E
2 C

3 B
4 D
5 A

A4

A 1
B 3

A5

1 Mirrors
2 Signal
3 Position
4 Speed
5 Look

A6

assess
decide
act

A7

A

A8

B

A9

D

A10

1 mirrors,
  position
2 signal
3 safe
4 safe
  distance
5 overtake

Crossroads
Question on page 28

A11

A 3 Crossroads. Priority f
traffic on the major road.
Never assume other driver
will give you priority.
B 1 Unmarked crossroad
C 2 Crossroads with give
way lines at the end of you
road. Give way to traffic on
the major road.

Roundabouts
Questions on pages 29 - 3

A12

C

A13

1 Left
2 Left
3 Right. Remember to
use the MSM routine befo
signalling left to turn off.

A14

1 Left
2 Going ahead
3 Right

A15

mirrors,
signal,
position,
speed,
look

A16

Part 4

Parking (on the road)
Questions on page 31

A1

1 safe
2 convenient
3 legal

A2

Cars 1,2,3,6

Part 5

Passing stationary vehicles and obstructions
Questions on pages 32 - 33

A1

2

A2

2, even though the obstruction is on the right. Where safe, when travelling downhill be prepared to give priority to vehicles (especially heavy vehicles) that are coming uphill.

Meeting and crossing the path of other vehicles
Questions on pages 34 - 35

A3

1 True
2 False. Always consider whether it is safe. Are there dangers the other driver cannot see? Remember, flashing headlamps has the same meaning as sounding the horn. it is a warning: 'I am here!' Sometimes it is taken to mean: 'I am here and I am letting you pass.'
3 True

A4

1 Yes
2 Yes
3 No
4 Yes
5 Yes
6 Yes
7 Yes

Part 6

Stopping in an emergency
Questions on pages 36 - 37

A1

1 True
2 True
3 True
4 False
5 False. Looking in the mirror should not be necessary. You should know what is behind you.
6 False
7 True
8 True

A2

pump
1 maximum
2 lock
3 quickly

A3

steer
brake
pump
pressure
1 False. Other elements beyond braking can cause skidding e.g. acceleration or going too fast into a bend.
2 False. Although you may stop in a shorter distance, you still need to leave the correct distance to allow yourself time to react and vehicles behind you time to stop.

Stopping distances
Questions on pages
38 - 39

A4

1 No
2 Yes
3 Yes
4 No

A5

1/2 second

A6

1 23 metres/75 feet
2 15 metres/50 feet
3 96 metres/315 feet

A7

longer
more

A8

breaks
two second

Part 7

Moving off at an angle
Question on page 40

A1

1 False. You should check your mirrors and blindspot before moving out. Keep alert for other traffic as you pull out and stop if necessary.
2 True
3 False. Move out slowly and carefully.
4 True. The closer you are, the greater the angle.
5 True. As you move out, you are likely to move on to the right-hand side of the road and into conflict with oncoming vehicles.
6 False. You should signal only if it helps or warns other road users. Signalling gives you no right to pull out.

Moving off uphill
Question on page 41

A2

1 True
2 False. Using the gas pedal will not move the car forwards.
3 False. As your feet will be using the clutch pedal and the gas pedal you need to use the handbrake to stop the car rolling back.
4 True
5 True .
6 True
7 True

Moving off downhill
Question on page 42

A3

1 True
2 False. Almost certainly you will need to use the foot-brake.
3 False. It is often better to move off in 2nd gear.
4 True. This will stop the car rolling forwards.
5 True
6 False. You will need to have your foot on the foot-brake to stop the car rolling forwards.

Approaching junctions uphill and downhill
Question on page 43

A4

The following statements are correct:
1,4,5,6,7

Part 8

Reversing
Questions on pages
44 - 45

A1

safe
convenient
law

A2

1 False
2 False
3 True
4 True

A3

A

A4

1,4

A5

A To the left
B To the right

A6

observation

Reversing into a side road on the left

Questions on pages 46 - 47

A7

3

A8

1 True
2 False
3 True

A9

4

A10

C

A11

Left

A12

The front of the car will swing out to the right.

A13

1 True
2 False
3 True

A14

pedestrians
road users
stop

Reversing into a side road on the right

Questions on pages 48 - 49

A15

True

A16

B

A17

2

A18

True. You may need to place your left hand at 12 o'clock and lower your right hand.

A19

True

A20

2

A21

B

Turning in the road

Questions on pages 50 - 51

A22

A

A23

slowly briskly

A24

False, but you should try to complete the manoeuvre in as few moves as possible.

A25

2,3,4

A26

all round

A27

1 Right
2 Steer briskly left
3 Left
4 Steer briskly right
5 Right

A28

1 Left
2 Over your right shoulder to where the car is going

Reverse parallel parking
Questions on pages 52 - 53

A29

2

A30

C , in line with the rear of the parked vehicle

A31

2

A32

1 To the left
2 The nearside headlamp of the vehicle towards which you are reversing
3 Clipping the rear offside of the lead car
4 Take off the left lock
5 Steer to the right and then take off the

right lock as you get straight

A33

1 False
2 False
3 True. You will be expected to be able to complete the exercise within approximately two car lengths.

A34

All round, particularly for pedestrians and oncoming vehicles

Part 9

Traffic lights and yellow box junctions
Questions on pages 54 - 55

A1

1 red
2 red and amber
3 green
4 amber
5 red

A2

1 green
2 red and amber
3 amber
4 red

A3

1 False
2 True
3 True
4 False.
Pedestrians who are already crossing have priority.

A4

2

A5

1 Lane A or B
2 Lane C
3 Lane A

A6

1 No
2 No ⎦ If your exit is blocked you should not enter a yellow box junction.
3 Yes

Pedestrian crossings
Questions on pages 56 - 57

A7

1 Zig-zag lines
2 Flashing yellow beacons on both sides of the road
3 Black and white stripes on the crossing
4 A give way line

A8

. True
. True.   You must
ot park or wait on the zig-zag
ines on either side of the
rossing.
. False.   You must not
vertake on the zig-zag lines on
pproach to the crossing.
. True
. True.   A slowing down arm
ignal should be used. It helps
edestrians understand what
ou intend to do. They cannot
ee your brake lights.

A9

. Traffic lights
. Zig-zag lines
. A white stop line

A10

. A white stick means
he pedestrian is visually
andicapped. A white stick with
wo red reflector bands means
he pedestrian may be deaf as
well as visually handicapped.

A11

. Flashing amber
. You must give way to
edestrians on the crossing,
out if it is clear you may go on.

A12

A bleeping tone. This sounds
when the red light shows to
drivers and helps visually

handicapped
pedestrians know
when it is safe to cross.

Level crossings
Questions on page 58

A13

A   3
B   1
C   2
D   4

A14

2
4
5

Part 10

One-way systems
Questions on page 59

A1

A   is the correct
sign for a one-way
street.
B   tells you 'Ahead
only'.

A2

1   True
2   True
3   True
4   True

Road markings
Questions on pages
60 - 61 ·

A3

information
warnings
orders

A4

1   They can be seen
when other signs
may be hidden
2   They give a
continuing message

A5

A   2
B   2

A6

1,4

A7

They are used
to separate
potentially dangerous
streams
of traffic.

A8

You must not enter
the hatched area.

Traffic signs
Questions on pages
62 - 63

A9

1 Warning
2 Order
3 Information

A10

1 must
2 must not

A11

1 You must give way to traffic on the major road. Delay your entry until it is safe to join the major road.
2 You must stop (even if the road is clear). Wait until you can enter the new road safely.

A12

vision is limited

A13

Motorway signs	D
Primary routes	E
Other routes	B
Local places	C
Tourist signs	A

Part 11

Road observation
Questions on pages 64 - 65

A1

speed
behaviour
intentions

A2

1 Observe that the view into the new road is restricted.
The driver should . . .
Move forward slowly, to get a better view.
Note the pedestrians who may walk in front of or behind car A.
Note the pedestrian waiting to cross.
Allow the cyclist to pass.
Once in a position to see car B, stop and give way.
2 Observe that the parked car restricts the view into and out of the side road.
The driver should . . .
Slow down on approach to parked car P.
Take up position to gain a better view and be more visible to car A and the pedestrian.
Slow down in case the pedestrian walks out from behind the parked car P.
Consider signal to pass parked car P.
Look carefully into minor road. Note the actions of car A. Be prepared to stop.

A3

bike

A4

3, but dabbing the brakes may encourage the driver to drop back

A5

A
1 Junctions
2 Hump-back bridges
3 Concealed entrances
4 Dead ground
5 Narrow lanes
B
1 Children playing
2 Horses
3 Pedestrians
4 Especially elderly and young cyclists
5 Other vehicles

On the open road (dealing with bends)
Questions on pages 66 - 67

A6

speed
gear
position

A7

1

A8

B

A9

A

A10

A   On a right-hand bend keep to the left. This will help to improve your view.
B   On a left-hand bend keep to the centre of the lane. Do not move to the centre of the road to get a better view. A vehicle travelling in the opposite direction may be taking the bend wide.

Overtaking
Questions on pages 68 -69

A11

necessary
1   Mirrors
2   Position
3   Speed
4   Look
5   Mirrors
6   Signal
7   Manoeuvre

A12

About the width of a small car, more in windy or poor weather conditions

A13

1   The vehicle in front is signalling and positioned to turn right
2   You are using the correct lane to turn left at a junction
3   Traffic is moving slowly in queues and the traffic on the right is moving more slowly than you are
4   You are in a one-way street

A14

1   On approach to a junction
2   The brow of a hill
3   The approach to a bend
4   Where there is dead ground   NB These are examples. Be guided by *The Highway Code*

Dual carriageways
Questions on pages 70 - 71

A15

2   Statements 1,3, 4 and 5 do not apply:
1   Reflective studs are used on some dual carriageways.
3   The speed limit is subject to local conditions and may vary from 40mph up to the national speed limit.
4   You can turn right on to and off dual carriageways

unlike motorways, where all traffic enters and leaves on the left.
5   You may find slow moving vehicles sometimes displaying a flashing amber light.

A16

A   You would cross over the first carriageway then wait in the gap in the central reservation. Be careful, if you are towing or if your vehicle is long, that you do not cause other road users to change course or slow down.
B   You would wait until there is a gap in the traffic long enough for you safely to clear the first carriageway and emerge into the second.

A17

The speed limit applies to all lanes. Use the first lane to travel in and the second for overtaking.

A18

A   Dual carriageway ends
B   Road narrows on both sides
C   Two-way traffic straight ahead

A19

A and C

Part 12

Driving an automatic car
Questions on pages 72 - 75

A1
Clutch

A2
1   Driving is easier
2   There is more time to concentrate on the road

A3
Park — Locks the transmission. This should be selected only when the vehicle is stationary.
Reverse — Enables the car to go backwards, as in a manual car.
Neutral — Has the same function as in a manual car. The engine is not in contact with the driving wheels.
Drive — Is used for driving forwards. It automatically selects the most appropriate gear.
3rd ⎤
2nd ⎥ Have the same
1st ⎦ function as manual gears

A4
3

A5
1   Yes
2   No
3   Yes, if you needed extra control
4   You would probably use kickdown, but possibly in certain circumstances you would manually select a lower gear
5   Yes, maybe using 1st gear
6   No. Use the brakes

A6
1   The right foot
2   The right foot

A7
It stops you trying to control the brake and accelerator at the same time. It encourages early release of the accelerator and progressive braking.

A8
Drive, reverse, all forward gears.

A9
1   You should apply the handbrake every time you stop. Otherwise you have to keep your foot on the foot-brake.

A10
Park (P) or Neutral (N)

A11
1

Part 13

The driving test
Questions on pages 76 - 81

A1
1   False
2   True
3   False
4   False
5   True

A2
3

A3
False. If you did not hear clearly or did not understand what the examiner said, you should ask him or her to repeat the instruction. If you have any problem with your hearing, it is advisable to tell the examiner at the start of the test.

A4

No. The standard of the test does not vary. The test result should be the same wherever it is taken.

A5

,4

A6

:

A7

The test will not proceed. You have failed not only the eyesight section, but the whole test. Remember, if you wear glasses or contact lenses, to wear them for the eyesight test and for the rest of the driving test.

A8

provisional licence, signed

A9

False. You can make some minor errors and still reach the required standard.

A10

False. You will be asked to stop your car as if in an emergency and two of the other three manoeuvres.

A11

Yes, but remember to do it up again when you have completed the exercise.

A12

1 Give you a verbal explanation of the main reasons for failure
2 Write out a form for you to take away showing you your main errors

A13

1 Drive unsupervised
2 Drive on a motorway
3 Drive without L-plates

A14

No. You must have had at least three years' driving experience (and be over 21 years of age).

A15

To the DVLA, Swansea

A16

Yes. It is a good idea to keep a note of your driver number and the date you passed your test.

A17

rules
high speed

A18

2 The examiner will expect you to drive normally. You should abide by all speed limits and drive according to road and traffic conditions.

A19

Yes. The examiner will be skilled in giving instructions and directions to deaf candidates.

Part 14

Beyond the test
Questions on pages 82 - 83

A1

No

A2

1 Bad weather driving
2 Night-time driving
3 Motorway driving
4 Skid control...and more

A3

2

A4

1, although people of any age can find it difficult to drive at night

A5

70

A6

There are certain serious driving offences which carry the penalty of disqualification. In order to regain a full licence, the disqualified driver has to apply for a provisional licence and take an extended test. If, because of certain illnesses, you have been unable to drive for 10 years, you will be required to take the test again in order to gain a full licence.

Motorway driving

Questions on pages 84 - 87

A7

1

A8

1 No
2 Yes
3 No
4 Yes
5 Yes
6 Yes
7 No

A9

1 Oil
2 Water
3 Fuel
4 Tyre pressures
These are some of the checks. You should take the advice given in the HMSO publication *Driving*.

A10

3 You should never attempt to retrieve anything from the carriageway.

A11

3

A12

B Amber
A Red
D Green
C White

A13

3

A14

Yes

A15

3

A16

1 False
2 False
3 False
4 True
5 False

A17

1, except when traffic is moving slowly in queues and the queue on the right is travelling more slowly

Safe night-time driving

Questions on pages 88 - 89

A18

2

A19

Switch on earlier, switch off later.

A20

1. It helps others to see you

A21

If you are stationary, to avoid danger from a moving vehicle.

A22

A. Always park with the flow of traffic. You will show red reflectors to vehicles travelling in your direction.

Part 17

A23

Pedestrians ⎤ two
Cyclists ⎬ of
Motor cyclists ⎦ these

A24

In poor weather
conditions See
and be seen

A25

dazzle
handbrake

All-weather
driving
Questions on pages
90 - 91

A26

1

A27

double

A28

aquaplaning

A29

2

A30

1

A31

very light

A32

2

A33

1   Slow down
2   stop
3   windscreen
wipers
4   demister, heated
rear windscreen

A34

100 metres/yards

A35

2

A36

2

A37

1   windows, audio
system.
Listen
2   early
3   brakes, alert
4   horn, warn

A38

False, because
your tyres are not in
contact with
the road

A39

brakes

A40

1   gently
2   lower gear
3   drop, brakes

A41

1

A42

If possible, control
your speed before
reaching the hill.
Select a low gear
early.

A43

Using your brakes

A44

Avoid harsh
acceleration

A45

1   The driver
2   The vehicle
3   The road
conditions

A46

1   Slowing
2   Speeding
3   Turning
4   uphill, downhill

Part 15

Vehicle care
Questions on pages
96 - 97

A1
1

A2
Oil

A3
3

A4
Water, anti-freeze, air

A5
1

A6
dazzle

A7
clean

A8
uneven, bulges, cuts

A9
2

A10
Get them checked as
quickly as possible

Breakdowns, accidents and emergencies
Questions on pages
98 - 103

A11
1   Neglect
2   routine checks
3   preventative
4   abuse

A12
1   50m/yds
2   At least 150m/yds

A13
Yes. Try to give as
much warning as
possible.

A14
3

A15
3

A16
Under the drawing of
the handset is an arrow
which points to the
nearest telephone.

A17
1   The emergency number
(painted on the box)
2   Vehicle details (make,
registration mark, colour)
3   Membership details of your
motoring organisation
4   Details of the fault

A18
1   By displaying a Help
pennant
2   By using a mobile
telephone

A19
1   passing motorists
2   do not know
3   leave, longer

A20
1   False.   Do not move
injured people unless they are
in danger.
2   False.   Tell them the
facts, not what you think is
wrong.
3   False.   Do not give those
injured anything to eat or
drink. Keep them warm and
reassure them.
4   False.   Keep hazard lights
on to warn other drivers.
5   True.   Switch off engines.
Put out cigarettes.
6   True,   in the case of injury.

A21
Stop

A22

No

A23

1   The other driver's name, address and contact number
2   The registration numbers of all the vehicles involved
3   The make of the other car
4   The other driver's insurance details
5   If the driver is not the owner, the owner's details

A24

1   Pull up quickly
2   Get all passengers out
3   Call assistance

A25

1   First aid
2   Fire extinguisher
3   Warning triangle

A26

3

The motor car and the environment
Questions on pages 104 - 105

A27

carbon dioxide, greenhouse

A28

4

A29

Yes

A30

False. A catalytic convertor reduces the level of carbon monoxide, nitrogen oxide and hydrocarbons by up to 90 per cent. Carbon dioxide is still produced.

A31

2

A32

2

A33

2

A34

Nine measures are listed here:
1   Make sure your vehicle is in good condition and regularly serviced.
2   Make sure tyres are correctly inflated. Under-inflated tyres waste fuel.

3   Push the choke in as soon as possible when starting from cold.
4   Avoid harsh braking.
5   Buy a fuel-efficient vehicle.
6   Use the most appropriate gear.
7   Use your accelerator sensibly and avoid harsh acceleration.
8   Use unleaded fuel.
9   Dispose of waste oil, old batteries and used tyres sensibly.

Buying a used car
Questions on pages 106 - 107

A35

dealer, auction, privately

A36

Why is it being sold?

A37

1   True
2   False
3   True
4   True

A38

1   What is covered
2   The length of the agreement

A39

taxed, MOT, insurance

A40

Four items to check
are listed here:
1 Mileage
2 Has it been
involved in any
accidents?
3 Number of owners
4 Is there any hire
purchase or finance
agreement
outstanding?

A41

True. The AA offer a
national inspection
scheme.

Part 16

Towing a caravan or trailer

Questions on pages
108 - 109

A1

1

A2

Exterior towing
mirrors, to give you a
good view

A3

2

A4

stabiliser

A5

3

A6

A   2
B   2

A7

Seven checks are listed here:
1 Is the caravan or trailer
loaded correctly?
2 Is it correctly hitched
up to your vehicle?
3 Are the lights and
indicators working properly?
4 Is the braking system
working correctly?
5 Is the jockey wheel
assembly fully retracted
and in the correct position?
6 Are tyre pressures
correct?
7 Are all windows, doors
and roof lights closed?

A8

jockey wheel, corner steadies

Driving on the Continent

Questions on page 110 - 111

A9

vehicle, documents

A10

1 route
2 motoring regulations

A11

Here are five routine
checks:
1 Tyres, including spare.
Always carry a spare tyre.
2 Tool kit and jack.
3 Lamps and brake lights.
4 Fit deflectors to your
headlampsto prevent dazzle
to other drivers approaching
on the left.
5 Check you have an extra
exterior mirror on the left.

A12

driving licence

A13

International Driving Permit

A14

Some non-EC countries

A15

2

A16

Five items are listed here:
1 Spare lamps and bulbs
2 Warning triangle
3 First aid kit
4 Fire extinguisher
5 Emergency windscreen